Introduction to General Chemistry

11ᵗʰ Hour

Introduction to General Chemistry

Stephen B. Barone

Aeronomy Laboratory
National Oceanic and Atmospheric Administration
and
The Cooperative Institute for Research in Environmental Sciences
University of Colorado
Boulder, Colorado

Blackwell
Science

© 2000 by Blackwell Science, Inc.

Editorial Offices:
Commerce Place, 350 Main Street, Malden, Massachusetts 02148, USA
Osney Mead, Oxford OX2 0EL, England
25 John Street, London WC1N 2BL, England
23 Ainslie Place, Edinburgh EH3 6AJ, Scotland
54 University Street, Carlton, Victoria 3053, Australia
Other Editorial Offices:
Blackwell Wissenschafts-Verlag GmbH, Kurfürstendamm 57, 10707 Berlin, Germany
Blackwell Science KK, MG Kodenmacho Building, 7-10 Kodenmacho Nihombashi, Chuo-ku, Tokyo 104, Japan

Distributors:
USA

    Blackwell Science, Inc.
    Commerce Place
    350 Main Street
    Malden, Massachusetts 02148
    (Telephone orders: 800-215-1000 or 781-388-8250; fax orders: 781-388-8270)

Canada

    Login Brothers Book Company
    324 Saulteaux Crescent
    Winnipeg, Manitoba, R3J 3T2
    (Telephone orders: 204-837-2987)

Australia

    Blackwell Science Pty, Ltd.
    54 University Street
    Carlton, Victoria 3053
    (Telephone orders: 03-9347-0300; fax orders: 03-9349-3016)

Outside North America and Australia
    Blackwell Science, Ltd.
    c/o Marston Book Services, Ltd.
    P.O. Box 269
    Abingdon
    Oxon OX14 4YN
    England
    (Telephone orders: 44-01235-465500; fax orders: 44-01235-465555)

Acquisitions: Nancy Hill-Whilton
Development: Jill Connor
Production: Louis C. Bruno, Jr.
Manufacturing: Lisa Flanagan
Marketing Manager: Carla Daves
Director of Marketing: Lisa Larsen
Interior Design by Colour Mark
Cover design by Madison Design
Typeset by Best-set Typesetter Ltd., Hong Kong
Printed and bound by Capital City Press

Printed in the United States of America
00 01 02 03 5 4 3 2 1

The Blackwell Science logo is a trade mark of Blackwell Science Ltd., registered at the United Kingdom Trade Marks Registry

Library of Congress Cataloging-in-Publication Data

Barone, Stephen B.
  General chemistry / by Stephen B. Barone.
    p. cm.—(11th hour)
  ISBN 0-632-04293-1
    1. Chemistry. I. Title. II. 11th hour (Malden, Mass.)
  QD33.B245 2000
  540—dc21   99-089911

CONTENTS

11TH HOUR GUIDE TO SUCCESS

The 11th Hour Series is designed to be used when the textbook doesn't make sense, the course content is tough, or when you just want a better grade in the course. It can be used from the beginning to the end of the course for best results or when cramming for exams. Both professors teaching the course and students who have taken it have reviewed this material to make sure it does what *you* need it to do. The material flows so that the process keeps your mind actively learning. The idea is to cut through the fluff, get to what you need to know, and then help you understand it.

Essential Background. We tell you what information you already need to know to comprehend the topic. You can then review or apply the appropriate concepts to conquer the new material.

Key Points. We highlight the key points of each topic, phrasing them as questions to engage active learning. A brief explanation of the topic follows the points.

Topic Tests. We immediately follow each topic with a brief test so that the topic is reinforced. This helps you prepare for the real thing.

Answers. Answers come right after the tests; but, we take it a step farther (that reinforcement thing again), we explain the answers.

Clinical Correlation or Application. It helps immeasurably to understand academic topics when they are presented in a clinical situation or an everyday, real-world example. We provide one in every chapter.

Demonstration Problem. Some science topics involve a lot of problem solving. Where it's helpful, we demonstrate a typical problem with step-by-step explanation.

Chapter Test. For more reinforcement, there is a test at the end of every chapter that covers all of the topics. The questions are essay, multiple choice, short answer, and true/false to give you plenty of practice and a chance to reinforce the material the way you find easiest. Answers are provided after the test.

Check Your Performance. After the chapter test we provide a performance check to help you spot your weak areas. You will then know if there is something you should look at once more.

Sample Midterms and Final Exams. Practice makes perfect so we give you plenty of opportunity to practice acing those tests.

The Web. Whenever you see this symbol ▣ the author has put something on the Web page that relates to that content. It could be a caution or a hint, an illustration or simply more explanation. You can access the appropriate page through *http://www.blackwellscience.com*. Then click on the title of this book.

The whole flow of this review guide is designed to keep you actively engaged in understanding the material. You'll get what you need fast, and you will reinforce it painlessly. Unfortunately, we can't take the exams for you!

PREFACE

The goal of this book is to provide chemistry students with additional resources useful in mastering the basic principles of chemistry. Chemistry occupies a unique position in the physical sciences in that it is an interfacial field uniting the diverse disciplines of biology, geology, and physics. A strong background in chemistry, therefore, is valuable to promoting success in a range of scientific studies and endeavors. A mastery of introductory chemistry unites a solid conceptual understanding of first principles with strong analytical skills for problem solving. This book is designed to promote this multifaceted understanding for students of a variety of backgrounds.

The structure of the text has several features to maximize its usefulness for a broad range of learners. The book is organized into thirteen chapters each of which is further broken down into central topics highlighting individual aspects of chemistry. This topic-oriented treatment maximizes its usefulness both as an introduction to new material and as a reference to aid in mastering concepts and skills needing more practice. Each chapter begins with a checklist of background concepts vital to the mastery of the new material presented. Students are strongly advised to refer to this checklist if initially concepts and problems are hard going. Following the checklist is a discussion of the essential concepts central to a given topic. This discussion draws on example problems, flow diagrams, and figures aimed at emphasizing the interconnected nature of the ideas presented. After each discussion, sample problems and solutions are provided to place the concepts in a practical, skill-building context. Because creative problem solving is essential for success in introductory chemistry, students are strongly advised to work through each problem presented in this format. Additional exercises to hone these important skills are provided in the demonstration problems, end of the chapter exams, and practice midterm and final exams. To provide additional depth and rigor, central aspects of the text, indicated by , are linked to an interactive web-based learning site located at www.blackwellscience.com.

Another purpose of the text is to provide examples of how chemistry is intimately connected to the world in which we live. Included in each chapter is an application that presents several of the concepts discussed in a real world context. These sections are intended to reinforce the material and develop interest in the subject. By promoting your own innate curiosity and interest in the material, the study of chemistry can be transformed into an enjoyable experience.

I gratefully acknowledge the following professors, and students for their manuscript reviews and helpful comments: Robley J. Light, Florida State University; Don Williams, Hope College; Kathleen Donnelly, Russell Sage College; Jim Long, University of Oregon; Kristen Kolberg, Russell Sage College; Jen MacLaughlin, Russell Sage College; Jenny Hansen, Ursinus College; and Christy Kenny, University of Pennsylvania School of Medicine. Sincere thanks go to Craig Benson of George Washington University for carefully reading the manuscript and verifying the topic tests and answers.

Stephen B. Barone
NOAA, Aeronomy Laboratory
Boulder, Colorado

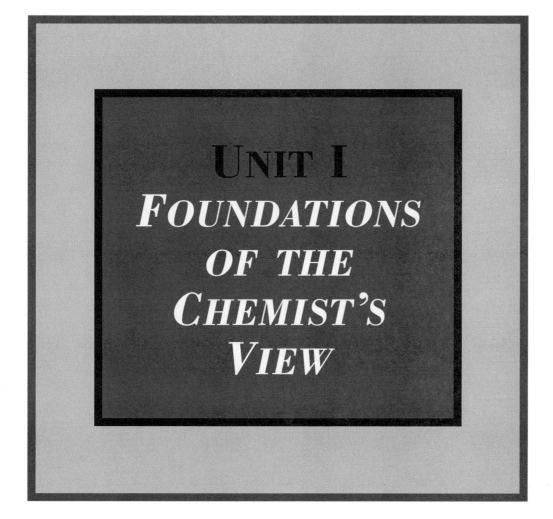

UNIT I
FOUNDATIONS OF THE CHEMIST'S VIEW

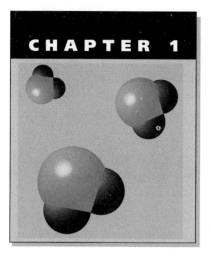

CHAPTER 1

Matter, Measurement, and the Scientific Method

Have you ever wondered how the universe began? Why life is sustained on Earth? Or even what factors control the formation of clouds on a sunny day? If you choose to explore any scientific subject, you will find that you are not alone in your curiosity. For thousands of years human beings have sought to understand the complex physical environment that surrounds us. The same curiosity that fueled ancient Chinese alchemists' pursuit of the elixir of life propels modern day scientists to develop faster computers, alternative sources of energy, and novel cures for disease. The study of chemistry is central to each of these endeavors and accordingly is strongly tied to our ability to perceive the changes that our physical environment undergo. Historically, chemists have sought to understand these changes in terms of the microscopic building blocks that comprise the physical world. This microscopic description has evolved over thousands of years through the development of scientific theories from experimentation, measurement, and the scientific method. As our ability to perceive the microscopic nature of matter expands, so too is our knowledge of the mysteries of the universe likely to rapidly grow.

ESSENTIAL BACKGROUND

- **Algebra and basic math skills**

- **Scientific notation**

- **SI (*le Systeme International d'Unites*)-based units**

TOPIC 1: PROPERTIES OF MATTER

KEY POINTS

✓ *What are the basic building blocks of matter?*

✓ *How do the three states of matter differ?*

✓ *What is the difference between a chemical change and a physical change?*

The study of chemistry seeks to explain the highly dynamic nature of the universe in terms of the fundamental properties and behavior of **matter**. In our context, matter is defined as the physical material that comprises the universe. In other words, matter is anything that possesses mass and occupies volume. This definition of matter is incredibly broad and includes everything from the cells that make up your body to the farthest moon of Saturn. In light of the diversity of material composing our universe, some means of classifying matter is needed if even the most basic understanding of nature is to be attained. Thousands of years of thought and experimentation have established that the unifying connection of all matter lies in its elemental composition. All matter consists of combinations of only about 100 basic building blocks that we designate as elements. It is this common thread that fuels the chemist's goal to explain the properties of matter in terms of its fundamental composition.

Elements are defined as substances that cannot be decomposed by chemical means into simpler substances. Elements differ from each other in their atomic composition. **Atoms** are the infinitesimally small building blocks that comprise elements. Each type of element is comprised of a unique kind of atom. Elements can be combined and arranged into specific shapes to form **compounds** or **molecules** by chemical reactions. Elements and compounds exist as **pure substances** if only one kind of matter is present. Alternatively, they can exist as **mixtures** that contain more than one kind of matter. Therefore, all matter may be viewed as originating from combinations of the following basic building blocks from simplest to most complex: atoms → elements → compounds.

Another way of categorizing matter is in terms of its physical state. Depending on physical conditions such as temperature or pressure, all matter exists in one of three states or phases: **solids, liquids,** and **gases**. To the maked eye, the three states of matter differ primarily in their rigidity or compressibility (e.g., solids are highly rigid, liquids are moderately rigid, and gases are nonrigid). On a microscopic level, the three states of matter differ by the strengths of the interactions that hold them together. For example, the water molecules that make up a cube of ice are held together much more tightly than those existing as a gas. Often, several states of matter are simultaneously present in the form of a **heterogeneous mixture**. Such mixtures are expected to have a variable composition in different regions of the mixture, which correspond to different states. In contrast, a mixture of substances that occupies only one state of matter is called a **homogeneous mixture**. These mixtures exhibit a constant composition throughout a sample. The flow diagram in **Figure 1.1** shows how the states of matter and basic building blocks can be used to classify all forms of matter we encounter.

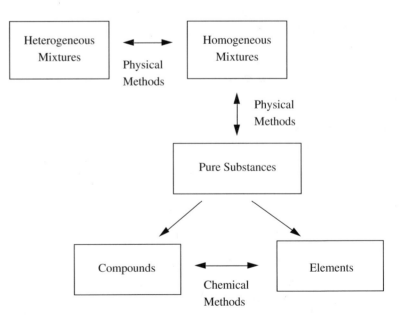

Figure 1.1. Classification of matter.

In the eyes of a chemist, the world is a dynamic environment driven by the many changes that matter undergoes. These changes are often viewed in terms of how they affect the physical state and chemical composition of matter. Changes that affect the state of matter without influencing the composition are termed **physical changes**. In contrast, changes that alter the chemical identity or composition of matter are designated **chemical changes**. For example, the evaporation of liquid octane (C_8H_{18}) in the piston of your automobile is a physical change: liquid → gas. However, the subsequent burning of gaseous octane to generate CO_2, H_2O, and energy is an example of a chemical change because the atoms in gasoline are rearranged into new forms.

Topic Test 1: Properties of Matter

True/False

1. Elements differ in the kinds of atoms that compose them. T

2. A carbonated beverage is an example of a homogeneous mixture. F

3. The condensation of water vapor to form a raindrop represents a chemical change. F

Multiple Choice

4. A solution prepared by dissolving table salt (NaCl) in water is an example of a(n)
   a. pure substance.
   b. element.
   c. homogeneous mixture.
   d. heterogeneous mixture.
   e. None of the above

5. Which of the following substances is an example of a pure substance?
   a. A diamond
   b. Carbon monoxide (CO)
   c. Liquid water
   d. Water vapor
   e. All of the above

Short Answer

6. A typical camping stove uses liquid butane (C_4H_{10}) as fuel for cooking. Consider the following processes that go into making a cup of instant coffee in the back country: 1. evaporate the liquid butane, 2. ignite (combust) a butane-air mixture, 3. boil water, and 4. dissolve instant coffee mixture. Explain each of these processes in terms of the physical/chemical changes that comprise them. 1) Physical change 2) chemical change 3) physical change 4) physical change

Topic Test 1: Answers

1. **True.** A specific element is composed of a unique kind of atom.

2. **False.** A carbonated beverage is a mixture of a liquid and gaseous CO_2 (hence, the bubbles). Therefore, it is a heterogeneous mixture (liquid + gas).

3. **False.** When a raindrop forms by condensation, water molecules change physical state (gas → liquid). Because no change of composition accompanies the change in state, the process is a physical change.

4. **c.** The mixture of dissolved NaCl and water is a homogeneous mixture because it consists of more than one pure substance (H_2O and NaCl) and occupies only one physical state (liquid).

5. **e.** Each choice is an example of a pure substance. a is an example of an element (carbon), and b through d are examples of pure compounds.

6. The evaporation of liquid butane (1) is a change of state (liquid → gas) and thus is a physical change. Combustion (2) involves the conversion of the compound butane (C_4H_{10}) to carbon dioxide (CO_2) and water (H_2O) and represents a chemical change. Boiling water (3) is a physical change in which liquid water is converted to vapor (liquid → gas). Dissolving instant coffee (4) is another example of a physical change, in this case involving a change of state from solid to liquid phases (solid → liquid).

TOPIC 2: SCIENTIFIC METHOD AND MEASUREMENT

KEY POINTS

✓ *What processes comprise the scientific method?*

✓ *What is the difference between a hypothesis and a theory?*

✓ *How is a measurement related to its unit?*

✓ *How can one system of units be converted into another?*

Chemistry is a highly dynamic science driven by observation. The dynamic nature of chemistry is found in the continuous testing and refining of scientific theories via experimentation. Over the last several centuries, a rigorous procedure governing the testing of scientific theories has developed, called the **scientific method**. The scientific method consists of four interconnected stages of inquiry: (1) observation and experiment, (2) hypothesis development, (3) prediction and experiment, and (4) hypothesis refinement and theory development. The method is shown schematically in **Figure 1.2** and is designed to use carefully conducted experiments to support or refute a scientific hypothesis. If a hypothesis withstands the test of a great number of experiments, it becomes known as a **theory**. The strength of the method lies in the continual testing and modification of hypotheses and theories by well-crafted experiments to test all aspects of the hypothesis in question. Therefore, it is the design and implementation of clever experiments that push forward the scientific frontier. A successful theory is widely supported by the results of a great number of experimental observations. Another consequence of the scientific method is the inability to prove any theory as scientific fact. The search for scientific truth must always be viewed as an ongoing one.

Scientific observations are often quantifiable by using some kind of measuring device, such as a thermometer or balance. In these cases, the measurement will always consist of a value and a unit of measure that reflect the absolute determination of the physical property in question. For example, if I were to report the measurement of the mass of a silicon computer chip to be 0.101, without specifying the unit of measure, the measurement would be meaningless. I would need to specify that the measurement in this case is in the unit of grams to relay useful information.

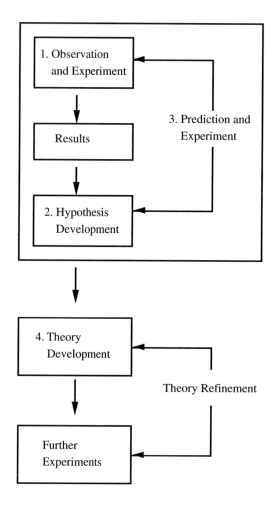

Figure 1.2. The scientific method.

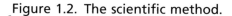

In addition to the value and unit of measure, most meaningful measurements also contain an estimate of the uncertainty associated with the value. This can be accomplished by attaching an error bar to the measurement (e.g., 0.101 ± 0.010) or by following the convention of significant figures. **Significant figures** are the digits in a measurement that are certain plus an additional final digit that is presumed to be uncertain. In this convention, this last digit is assumed to be accurate to a factor of ±1. For example, if the mass of a textbook is reported as 2.3 kilograms, an error estimate of 2.3 ± 0.1 can be inferred from the significant figures reported. Nonzero integers are always presumed to be significant. Zeros are considered significant if they lie between nonzero digits or if they follow nonzero integers and the number contains a decimal point. Thus, the numbers 1,001 and 11.00 each possess four significant figures. In addition to numbers with a finite number of significant figures, numbers that are known exactly are also encountered in scientific calculations. These numbers are called **exact numbers** and are presumed to possess an infinite number of significant figures. Such numbers commonly arise when defining units or when counting. Examples of exact numbers include physical constants such as the speed of light (c) or the numbers in the equality statement 4 quarts = 1 gallon. The rules for determining the number of significant figures are summarized in **Table 1.1**.

In many cases, several measurements are combined in a scientific calculation. Uncertainty in such calculations reflects both the number of significant figures in the measurements and the mathematical operations carried out. **Table 1.2** summarizes a useful set of guidelines for determining the number of significant figures when performing mathematical calculations. As an

Table 1.1 Rules for Determining the Number of Significant Figures

1. **Nonzero integers:** Nonzero digits are alway considered significant. The number 2.54 possesses three significant figures.
2. **Zeros**
   a. Leading zeros precede nonzero digits and are not considered to be significant. The number 0.000254 has three significant figures.
   b. Trailing zeros follow nonzero digits. They are only considered significant if the number has a decimal point. The number 2.540 has four significant figures.
   c. Captive zeros are zeros between nonzero digits. They always count as significant figures. The number 2.5004 has five significant figures.
3. **Exact numbers:** Numbers that are obtained by counting or by defining a unit system are called exact numbers and are assumed to possess an infinite number of significant figures. Examples include numbers in unit equality statements and some physical constants.

Table 1.2 Rules for Significant Figures in Calculations

1. **Multiplication and division:** When multiplying or dividing two or more numbers, the result possesses the same number of significant figures as the measurement with the least number of significant figures:

$$99.10 \times 1.4 = 140$$

2. **Addition and subtraction:** When adding or subtracting two or more numbers, the result possesses the same number of decimal places as the measurement with the least number of decimal places:

$$1.243 + 2.3156 + 1.20 = 4.76$$

3. **Logarithms:** When taking the logarithm of a number, the number of digits after the decimal place in the result is equal to the number of significant figures in the original number:

$$\text{Log}(79.1) = 1.898$$

example of the application of these rules, consider the following measurements to determine the density of an unknown metal sample: mass = 3,791 kilograms and volume = $4.5 \times 10^5 \text{cm}^3$. These measurements can be combined to yield a value of density certain to two significant figures: density = mass/volume = $3,791 \text{kg}/4.5 \times 10^5 \text{cm}^3 = 0.084 \text{kg/cm}^3$. Careful attention to significant figures is important in any scientific calculation.

The fundamental physical quantities we deal with most in our description of chemistry are mass, length, time, temperature, amount, and electric current. All other physical quantities such as volume, density, or pressure may be derived from combinations of these fundamental physical quantities. Most scientists exclusively use a common system of units when reporting scientific observations called the ***le Systeme International*** or the **SI system**. A convenient aspect of the SI system of units is the use of prefixes to change the size of the unit when dealing with measurements over a wide range of magnitudes. **Table 1.3** summarizes the most common of these prefixes. For example, a typical atom has a diameter of approximately 2×10^{-10} m. Using SI prefixes, this value can be expressed as 200 pm in units, which better reflects the incredibly small size of atoms. Converting a measurement in SI units to another unit system can be accomplished following the method of **dimensional analysis**. In this method, a **conversion factor** is used to change from one unit to another. A conversion factor is a fraction whose denominator and numerator are the same quantity expressed in different units. Accordingly, a measurement can be multiplied by the appropriate conversion factor without changing its value. For example, the conversion factor (2.54 cm/1 inch) can be used to convert a measurement of 4.72 inches to centimeters via the following calculation:

$$4.72 \text{ inches} \times (2.54 \text{ cm}/1 \text{ inch}) = 12.0 \text{ cm}$$

| Table 1.3 Common Prefixes Used in the SI System of Units |||
PREFIX	SYMBOL	MULTIPLE
Penta	P	10^{15}
Tera	T	10^{12}
Giga	G	10^{9}
Mega	M	10^{6}
Kilo	k	10^{3}
Hecto	h	10^{2}
Deka	da	10
Deci	d	10^{-1}
Centi	c	10^{-2}
Milli	m	10^{-3}
Micro	μ	10^{-6}
Nano	n	10^{-9}
Pico	p	10^{-12}

It should be noted that the reciprocal of a conversion factor is itself a useful conversion factor for the reverse unit conversion. For example, the 12.0 cm determined above can easily be converted back to inches via the following calculation:

$$12.0\,cm \times (1\ inch/2.54\,cm) = 4.72\ inches$$

Careful attention to dimensional analysis is essential to ensure that the solutions to complicated chemistry problems have the proper units and provides clues of potential errors in calculation.

Topic Test 2: Scientific Method and Measurement

True/False

1. Albert Einstein's theory of relativity is a scientific truth because it has never been disproved.   F

2. The SI unit of density [$\equiv$ (mass/volume)] is kg/m^3.   T

3. There are 1×10^6 seconds in a microsecond.   F

Multiple Choice

4. Any scientific measurement always
   a. contains a value and a unit.
   b. proves a scientific hypothesis.
   c. contains a measure of uncertainty.
   d. disproves a scientific hypothesis.
   e. None of the above

5. A large bathtub can hold 25 gallons of water. To how many milliliters does this amount correspond? (1 gallon = 3.7854 liters and 1,000 milliliters = 1 liter)
   a. 9.5×10^4 milliliters
   b. 6.6×10^{-3} milliliters

ok

$25 \times (3.7854\ y_1) = 94.635 \qquad 10^3 = 1\ L$

9.5×10^4

c. 6.6×10^3 milliliters

d. 9.5×10^{-2} milliliters

e. None of the above

Short Answer

6. An industrial accident releases 3,510 liters of unrefined petroleum into the ocean. If the density of unrefined petroleum is $0.912 \, \text{g cm}^{-3}$, what mass of petroleum was released in the accident? [Note: density = (mass/volume) and $1,000 \, \text{cm}^3 = 1$ liter.]

Topic Test 2: Answers

1. **False.** A scientific theory can never be proven, only disproved.

2. **True.** Density is an example of a quantity with units, kg/cm^3, derived from the fundamental SI units. Because density is defined as the mass per unit volume, it follows that the SI units would reflect the equation density = (mass [kg]/volume [m³]).

3. **False.** The metric prefix micro designates a factor of 1×10^{-6}. Therefore, there are 1×10^{-6} seconds in 1 microsecond.

4. **a.** Every measurement must have both a value (number) and unit to have physical significance.

5. **a.** The conversion factors (3.7854 liters/1 gallon and 1,000 milliliters/1 liter) can be used in the following calculation to determine the volume in milliliters:

$$(25 \text{ gallons}) \times (3.7854 \text{ liters/1 gallon}) \times (1{,}000 \text{ milliliters/1 liter}) = 9.5 \times 10^4 \text{ milliliters}$$

Notice that the calculation used the product of two conversion factors because we performed the dimensional analysis in two steps (gallon → liter → milliliters).

6. 3.20×10^6 grams. This short answer problem combines using the derived unit of density with dimensional analysis. Mass is related to density and volume by the following algebra:

definition of density: density = (mass)/(volume)

rearranging equation: mass = (density) × (volume)

solving for mass:
mass = $(0.912 \, \text{g/cm}^3) \times (3{,}510 \text{ liters}) \times (1{,}000 \, \text{cm}^3/1 \text{ liter})$
mass = $3.20 \times 10^6 \, \text{g}$

DEMONSTRATION PROBLEM

A cleanup crew at a local superfund site discovers 5.12×10^8 grams of carbon tetrachloride (CCl_4). If the density of CCl_4 is $1.6 \times 10^3 \, \text{g/L}$, what volume container, in units of kiloliters, is needed to remove the sample?

Solution

Use the density and mass of the CCl_4 sample to compute volume. Next, use the proper conversion factor to change units from liters to kiloliters.

definition of density: density = (mass)/(volume)

rearranging equation: volume = (mass)/(density)

solving for volume: volume = $(5.12 \times 10^8 \text{ grams})/(1.6 \times 10^3 \text{ g/L}) = 3.2 \times 10^5$ liters

converting to kiloliters: volume = $(3.2 \times 10^5 \text{ liters}) \times (1 \text{ kiloliters}/1{,}000 \text{ liters})$
$= 3.2 \times 10^2$ kiloliters

Chapter Test

True/False

1. Heterogeneous mixtures are composed of several pure substances. *T*

2. A theory is a hypothesis supported by a large number of experiments. *T*

3. Every chemical change must be accompanied by a change in physical state. *F*

4. There are exactly 1×10^9 nanoseconds in 1 second. *T*

5. The dissolution of glucose into the bloodstream is an example of a chemical change. *F*

Multiple Choice

6. The two nitrogen atoms in N_2 are separated by a distance of 2.12 angstroms (Å). Express this distance in meters given the following conversion factor ($1\text{Å} = 1 \times 10^{-8}\text{cm}$):
   a. 2.12×10^{10} meters
   b. 2.12×10^{-10} meters ✓
   c. 2×10^{-10} meters
   d. 2×10^8 meters
   e. 2.12×10^{-8} meters

7. Which of the substances below is an example of a homogeneous mixture?
   a. The human body
   b. Pure water ice
   c. Gasoline ✓
   d. A copper atom
   e. None of the above

8. For a scientific hypothesis to become a theory, it must
   a. be proven to be a scientific fact.
   b. be consistent with the theory that all matter is made of atoms.
   c. be an important scientific finding.
   d. not be disproved by an experiment to date. ✓
   e. None of the above

9. A 4.0-gram sample of helium is found to occupy a 22.4-liter volume at 25°C. What is the density of the helium sample at 25°C? $D = \dfrac{M}{V}$
   a. 0.18 g/L ✓
   b. 5.6 g/L
   c. 90 g/L
   d. 0.022 g/L
   e. 2.2 g/L

Short Answer

10. A French friend wants to run a 26-mile road race. How many kilometers should I tell him the race is (1 mile = 1.61 km)? 42

11. In a 1-cm^3 sample of room air there are 2.1×10^{19} molecules per cm^3. How many molecules would be in a cubic inch sample of air (2.54 cm = 1 inch)?

12. The speed of light is 2.9979×10^8 m/sec. To how many kilometers per hour does this value correspond?

Chapter Test Answers

1. **True**

2. **True**

3. **False**

4. **True**

5. **False**

6. **b** 7. **c** 8. **d** 9. **a**

10. 42 km

11. 3.4×10^{20} molecules per inch^3

12. 1.0792×10^9 km/hr

Check Your Performance

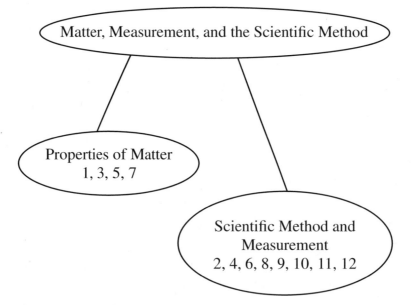

Use this chart to identify weak areas, based on the question numbers you answered incorrectly in the chapter test.

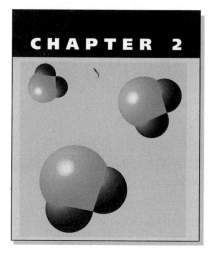

CHAPTER 2

Atoms, Molecules, and Ions

In light of the extraordinary diversity of substances that comprise our world, it is astounding to think that all matter is formed from combinations of a relatively small set of 112 elements. The 112 known elements are analogous to paints in a palette that can be used to create a seemingly infinite number of different finished canvasses. To develop an understanding of the chemical properties of matter, we must first understand the fundamental nature of the atoms that compose the known elements. In this chapter we examine the composition and properties of atoms and consider what factors govern their role in forming molecules and ions.

ESSENTIAL BACKGROUND

- **Dimensional analysis (Chapter 1)**
- **SI system of units (Chapter 1)**
- **The periodic table**

TOPIC 1: MODERN ATOMIC THEORY

KEY POINTS

✓ *What subatomic particles make up an atom?*

✓ *How do elements differ from one another?*

✓ *How do the isotopes of a given element differ?*

✓ *What are atomic mass and atomic number?*

Over the last two centuries a remarkably detailed description of the atom has evolved from the work of a great number of theorists and experimentalists. Clever deductions from scientists such as John Dalton, J.J. Thompson, and Ernest Rutherford were able to predict a great deal about the composition of atoms before technology allowed us to actually detect them. Since that time, huge strides in our understanding of atomic composition have been made. We focus our discussion on the current understanding of atomic structure as it applies to chemistry.

Atoms are composed of three subatomic particles: **protons, neutrons,** and **electrons**. Positively charged protons with a charge equal to $+1.602 \times 10^{-19}$ C and neutrons that possess no charge are located in a **nucleus** that occupies an incredibly small fraction of the total volume of an atom (nuclear diameter $\cong 10^{-13}$ cm). Electrons carry an equal charge in magnitude but oppo-

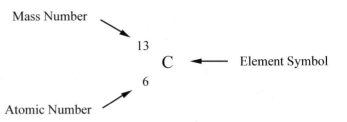

Figure 2.1. Atomic symbol representation of carbon-13.

site sign to that of a proton and move around the nucleus at an average distance of approximately 10^{-8} cm from its center. Although it occupies a very small volume, the nucleus provides almost the entire mass of an atom. The masses of protons and neutrons are nearly equal and are often expressed in terms of atomic mass units (mass of 1 proton $\cong$ mass of 1 neutron $\cong$ 1.0 amu). Although electrons possess almost no mass, it is their presence that gives rise to all the chemical properties of an atom.

Atoms differ from each other in the number of subatomic particles that compose them. Because atoms possess no net charge, the number of protons in a particular atom is always equal to the number of electrons. This number is designated as the **atomic number** and characterizes the chemical properties of the atom. The atomic number also identifies which element is associated with a particular atomic structure. The number of neutrons can vary from the atomic number and gives rise to isotopes of a particular atom. **Isotopes** are atoms that possess the same atomic number but different atomic masses. Therefore, the isotopes of an atom are differentiated from each other by their **atomic mass (atomic mass = number of protons + number of neutrons)**. Atomic number, atomic mass, and elemental identity are each summarized in the **atomic symbol** representation of an element. For example, the atomic symbol shown in **Figure 2.1** represents a carbon-13 atom possessing six protons, six electrons, and seven neutrons. All the known elements are organized in order of increasing atomic number in the **periodic table**. This ordering is extremely useful in explaining the chemical properties of a particular element because it reflects the number and type of electrons present that in turn determines its chemical reactivity.

Topic Test 1: Modern Atomic Theory

True/False

1. The mass of an atom is located almost entirely in the nucleus.

2. The number of neutrons in an atom determines its chemical properties.

3. The periodic table orders elements by increasing atomic mass.

Multiple Choice

4. Which of the following atomic symbols is incorrect?
   a. $^{81}_{35}Br$
   b. $^{14}_{6}C$
   d. $^{32}_{17}S$
   e. $^{16}_{8}O$

5. Which of the following pairs of atoms are isotopes of the same element?
   a. $^{126}_{53}I$ and $^{127}_{53}I$

b. $^{126}_{53}I$ and $^{126}_{52}Te$

c. $^{126}_{53}I$ and $^{127}_{54}Xe$

d. $^{12}_{6}N$ and $^{13}_{6}N$

e. None of the above

6. Any given atom has equal numbers of
   a. protons and neutrons.
   b. protons and electrons.
   c. neutrons and electrons.
   d. nuclei and electrons.
   e. atomic mass units.

Short Answer

7. The isotopes of carbon are used to date artifacts by radiocarbon dating techniques. Write out the atomic symbols for the isotopes of carbon corresponding to the atomic masses of 11, 12, 13, and 14 amu.

Topic Test 1: Answers

1. **True.** Protons and neutrons provide most of an atom's mass and are located in the nucleus. Electrons contribute negligibly to atomic mass.

2. **False.** The chemical properties of an atom are determined by the number and type of electrons. This is because electrons reside far from the nucleus and, thus, interact with other elements first and foremost.

3. **False.** The periodic table is arranged in order of increasing atomic number. Arranging the periodic table in this way best reflects the chemical properties of the elements.

4. **d.** Sulfur has an atomic number of 16. Therefore, the atomic symbol for a sulfur-32 isotope is $^{32}_{16}S$.

5. **a.** Any pair of isotopes will always have the same atomic number and different atomic masses. In this example, $^{126}_{53}I$ and $^{127}_{53}I$ is the only pair that satisfies this requirement.

6. **b.** Atoms are electrically neutral. Because protons and electrons have equal but opposite charges, the number of protons must always equal the number of electrons.

7. $^{11}_{6}C$, $^{12}_{6}C$, $^{13}_{6}C$, and $^{14}_{6}C$. The isotopes of a given element always possess identical atomic numbers but different atomic masses.

TOPIC 2: MOLECULES AND IONS

KEY POINTS

✓ *What are molecules?*

✓ *How do molecular and structural formulae differ?*

✓ *What is an ion? How is the charge of an ion predicted?*

✓ *How is an ionic compound formed?*

A **molecule** is an assembly of two or more atoms that are chemically bonded to each other. The atoms that comprise a particular molecule are held together in a specific arrangement by attractive forces called **chemical bonds**. As a result of these forces, a molecule can be viewed as a single distinct object with a definite chemical composition. Molecular composition reflects the number and type of atoms that are bound together and is conveniently expressed by the compound's **molecular formula**. For example, water has a molecular formula of H_2O, which indicates that a molecule of water contains two hydrogen atoms and one oxygen atom. The sum of atomic masses of each atom given in the molecular formula determines the **molecular mass** of a compound. For example, the molecular mass of H_2O is calculated in the following manner:

$$\text{Mass } H_2O = 1\,amu(\text{mass H}) + 16\,amu(\text{mass O}) + 1\,amu(\text{mass H}) = 18\,amu$$

The arrangement of atoms in a molecule can be explicitly shown in that compound's **structural formula**. For example, the structural formula: **H—O—H** shows that water consists of two hydrogen atoms chemically bonded to an oxygen atom.

Although atoms and molecules are electrically neutral species, certain processes can change the net number of electrons associated with them. These processes have no effect on the number of protons and neutrons in a given atom. However, they lead to the formation of electrically charged species call **ions**. A **cation** is a positively charged ion formed via the loss of one or more electrons. Conversely, an **anion** is a negatively charged ion formed by the addition of one or more electrons. The magnitude of the charge of an ion is calculated by subtracting the number of electrons present in the ion from the atom's atomic number: **charge = atomic number – number of electrons**. Typically, an ion will assume a net charge that enables it to achieve the same number of electrons as the closest noble gas, which are the elements composing the last column in the periodic table. **Polyatomic ions** are composed of several atoms joined together with a net positive or negative charge.

Cations and anions are attracted to each other because they possess opposite charges. As a result of this attraction, cation–anion pairs may interact together to form an **ionic bond** between them. The resulting **ionic compound** reflects the number and type of cations and anions that are bonded. Because ionic compounds are electronically neutral, anions and cations add to each other such that the total negative charge equals the total positive charge. For example, two Br^- anions, each with a -1 charge, associate with one Ca^{2+} cation to form the ionic compound $CaBr_2$.

Topic Test 2: Molecules and Ions

True/False

1. All molecules are composed of cation and anion pairs.

2. The molecular mass of CO_2 is 44 amu.

3. Cl^- can be formed from Cl via the loss of one electron.

Multiple Choice

4. Which of the following ions possess 53 protons and 54 electrons?
   a. Cl^-
   b. Te^{2+}
   c. He^{2+}

d. I$^-$

e. Ba^{2+}

5. Ca^{2+} and SO$_4^{2-}$ combined to form what ionic compound?
    a. CaSO$_4$
    b. Ca$_2$SO$_4$
    c. Ca(SO$_4$)$_2$
    d. Ca(SO$_4$)$_4$
    e. None of the above

6. How many oxygen atoms are present in the ionic compound Al(NO$_3$)$_3$?
    a. 1
    b. 3
    c. 6
    d. 9
    e. 12

Short Answer

7. Glyceryl trimyristate (C$_{45}$H$_{86}$O$_6$) is a molecule similar in structure to fat in human tissue. Give the number of carbon, hydrogen, and oxygen atoms that make up this molecule and calculate its molecular mass.

Topic Test 2: Answers

1. **False.** Molecules are composed of groups of atoms held together by chemical bonds. Ionic compounds are formed from the attraction of cation and anion pairs.

2. **True.** CO$_2$ is made up of one carbon atom and two oxygen atoms. The molecular mass is equal to the sum of the atomic masses: molecular mass = 12 amu [C] + 16 amu [O] + 16 amu [O] = 44 amu.

3. **False.** Cl$^-$ can be formed from Cl via the addition of an electron: Cl + e$^-$ → Cl$^-$.

4. **d.** The number of protons and electrons in I$^-$ can be determined from the atomic number and ionic charge. The number of protons equals the atomic number (53) and the number of electrons is the atomic number minus the value of the charge: 53 − (−1) = 54.

5. **a.** The anion and cation pair in an ionic compound come together such that the total negative charge equals the total positive charge (i.e., net charge equal to zero). Because Ca^{2+} and SO$_4^{2-}$ possess the same magnitude of charge, they will add in a one-to-one fashion.

6. **d.** Al(NO$_3$)$_3$ represents three NO$_3^-$ anions ionically bound to one Al^{3+} cation. Therefore, 3 NO$_3^-$ can be thought of contributing nine oxygen atoms to the ionic compound.

7. 722 amu. The subscripts in the molecular formula C$_{45}$H$_{86}$O$_6$ designate the number of each type of atom present in glyceryl trimyristate: 45 carbon atoms, 86 hydrogen atoms, and 6 oxygen atoms. The molecular mass of glyceryl trimyristate is the sum of the atomic masses represented in the molecular formula

$$\text{Molecular mass} = 45 \times (\text{mass of C}) + 86 \times (\text{mass of H}) + 6 \times (\text{mass of O})$$
$$= 540\,\text{amu} + 86\,\text{amu} + 96\,\text{amu} = 722\,\text{amu}$$

TOPIC 3: CHEMICAL NOMENCLATURE

KEY POINTS

✓ *Which compounds are classified as inorganic compounds?*

✓ *How are ionic compounds named?*

✓ *What is an acid? How are simple acids named?*

✓ *What are binary molecular compounds? How are they named?*

Over 20 million different chemical compounds have been identified by scientists to date. To provide some reasonable system of dealing with the sheer magnitude of known compounds, chemists have developed a formal set of procedures for naming them. The basis of chemical nomenclature involves ordering compounds into groups based on their chemical properties. Within each group, a set of generic rules exists for assigning names to compounds. Therefore, naming common compounds combines the identification of the correct class of molecule with the application of the proper set of rules to determine its name. All chemical compounds can be divided into two basic groups, organic compounds and inorganic compounds. Organic compounds contain carbon in some combination with oxygen, sulfur, hydrogen, and/or nitrogen. All other compounds are designated as inorganic compounds. We focus our discussion of nomenclature on three types of inorganic compounds: **ionic compounds, acids,** and **binary covalent compounds**.

Ionic compounds are named by naming the individual ions in the order of cation to anion: ionic compound = cation name + anion name. Monatomic cations assume the same name as their parent element. For example, Ca^{2+} is called calcium. If the monatomic cation can commonly form more than one charged state, a roman numeral indicating the magnitude of the positive charge follows the cation name in parentheses. For example, Cr^{2+} is named the chromium (II) ion. Monatomic anions are named by combining the root of the parent element name with the ending -*ide* (e.g., O^{2-} is ox*ide*). Ionic compounds are also formed by the attraction of polyatomic ions. **Table 2.1** summarizes the names and formulae of a number of important polyatomic anions. Although these names should be memorized, several general rules exist for naming polyatomic atoms. Polyatomic cations have specific names that generally end in -*ium* (e.g., NH_4^+,

Table 2.1 Names of Common Polyatomic Ions

NAME	FORMULA	NAME	FORMULA
Mercury (I)	Hg_2^{2+}	Nitrite	NO_2^-
Ammonium	NH_4^+	Nitrate	NO_3^-
Acetate	$C_2H_3O_2^-$	Peroxide	O_2^{2-}
Carbonate	CO_3^{2-}	Hydroxide	OH^-
Hydrogen carbonate	HCO_3^-	Sulfite	SO_3^{2-}
Hypochlorite	ClO^-	Sufate	SO_4^{2-}
Chlorite	ClO_2^-	Hydrogen sulfite	HSO_3^-
Chlorate	ClO_3^-	Hydrogen sulfate	HSO_4^-
Perchlorate	ClO_4^-	Thiosulfate	$S_2O_3^{2-}$
Cyanide	CN^-	Phosphate	PO_4^{3-}
Chromate	CrO_4^{2-}	Monohydrogen phosphate	HPO_4^{2-}
Dichromate	$Cr_2O_7^{2-}$		
Permanganate	MnO_4^-	Dihydrogen phosphate	$H_2PO_4^-$

ammon*ium*). Polyatomic anions containing oxygen are called oxyanions and generally end in *-ate* or *-ite*. The ending *-ate* is used for the most common oxyanion and *-ite* is used to designate an anion of same charge but one fewer oxygen atom. Similarly, the prefix *per-* is given to an oxyanion with one more oxygen than the most common form and the prefix *hypo-* is used to indicate an oxyanion with one less oxygen atom than the *-ite* form. Together, these basic rules outline anion "families" such as the one below:

$$ClO_4^-: \textbf{\textit{per}}chlor\textbf{\textit{ate}} \quad ClO_3^-: chlor\textbf{\textit{ate}} \quad ClO_2^-: chlor\textbf{\textit{ite}} \quad ClO^-: \textbf{\textit{hypo}}chlor\textbf{\textit{ite}}.$$

Ammonium perchlorate is an example of an ionic compound composed of two polyatomic ions, which has the formula NH_4ClO_4. The rules for naming simple ionic compounds are summarized in **Table 2.2**.

Acids are an important class of compounds that are discussed in detail in subsequent chapters. However, for the moment we will define an acid as a compound that produces H^+ cations and anions when dissolved in aqueous solution. This definition will be substantially refined in Chapter 12. For the purpose of naming, an acid may be considered one or more H^+ cations bound to an anion. The name of any acid is constructed using the name of the anion it possesses. If the anion name ends in *-ide*, the acid assumes a name consistent with the form *hydro-* + anion root + *-ic* acid. An example of this type of acid is hydrobromic acid given by the formula HBr. If the anion ends in *-ate*, the acid name will be constructed by combining the anion root with the ending *-ic*. For example, HNO_3 is called nitric acid. Analogously, anions ending in *-ite* are named combining the anion root with the ending *-ous*. For example, HNO_2 is called nitrous acid. The rules for naming acids are summarized in Table 2.2.

Binary covalent compounds form by the association of two nonmetal elements. To name binary covalent molecules, the element farthest to the left on the periodic table is written first followed

---

Table 2.2 Rules for Naming Simple Inorganic Compounds

Ionic compounds
    formula =   _____   _____
                     (Cation name)   (Anion name)

    Naming cation: 1. Element name from periodic table.
                      2. Specify ion charge with roman numeral if more than one charge possible.
                      3. Polyatomic cations have special names.
    Naming anion: 1. Root of element name + *-ide*.
                      2. Polyatomic anions have special names (memorize these!).

Acids: HX → H⁺ (cation) + X⁻ (anion)
    No oxygen present:
        formula = hydro_____ic acid
                     (root of anion name)
    Oxygen present:
        anion name ends in *-ate*:
            formula = hydro_____ic acid
                       (root of anion name)
        anion name ends in *-ite*:
            formula = hydro_____ous acid
                       (root of anion name)

Binary covalent compounds (two nonmetal elements)
    formula = _____ + _____ _____ + _____
             (prefix)   (first element)   (prefix)   (second element)
    1. Prefix denotes number of atoms.
    2. Never use prefix *mono-* for first element.

Table 2.3 Prefixes Used in Chemical Nomenclature	
NUMBER	GREEK PREFIX
1	Mono
2	Di
3	Tri
4	Tetra
5	Penta
6	Hexa
7	Hepta
8	Octa
9	Nona
10	Deca

by the second element. The first element is given its full element name and the second element is named as if it was a monatomic anion: root + *-ide*. Finally, Greek prefixes are used to indicate the number of each kind of atom present. These prefixes are summarized in **Table 2.3**. For example, Cl_2O is called *dichlorine monoxide*. The basic method for naming binary covalent compounds is also summarized in Table 2.2.

Topic Test 3: Chemical Nomenclature

True/False

1. Binary inorganic compounds produce H^+ cations when dissolved in water.

2. The correct name for Hg_2^{2+} is the mercury (I) ion.

3. Potassium sulfate is represented by the molecular formula KSO_4.

Multiple Choice

4. The chemical name of KNO_2 is
   a. potassium nitrate.
   b. potassium dioxide nitrate.
   c. nitrous acid.
   d. potassium nitride.
   e. potassium nitrite.

5. Which molecular formula represents dinitrogen pentaoxide?
   a. N_2O_5
   b. NO_3
   c. HNO_3
   d. N_5O_2
   e. NO_5

6. Which of the molecules below is an acid?
   a. NO_3^-
   b. Na_2SO_4
   c. H_2S

d. N_2O

e. KBr

Short Answer

7. Give the class of compound and name of the following compounds: NF_3, $Fe(NO_3)_3$, HSO_3, CCl_4, and NaOH.

Topic Test 3: Answers

1. **False.** Acids generate H^+ cation and an anion when dissolved in water.

2. **True.** Hg_2^{2+} is a cation in which each mercury atom carries a +1 charge.

3. **False.** Potassium sulfate has the formula K_2SO_4. Two K^+ cations are needed to balance the −2 charge of the SO_4^{2-} anion.

4. **e.** KNO_2 is an ionic compound. It is named by combining the cation name (potassium) with the anion name (nitrite).

5. **a.** N_2O_5 is a binary covalent compound. It is named by combing the prefixed first nonmetal element (dinitrogen) with the second prefixed nonmetal (pentaoxide).

6. **c.** H_2S is hydrosulfuric acid. In solution it generates H^+ and HS^- ions.

7. NF_3, binary covalent compound, nitrogen trifluoride; $Fe(NO_3)_3$, ionic compound, iron(III) nitrate; HSO_3, acid, sulfurous acid; CCl_4, binary covalent compound, carbon tetrachloride; and NaOH, ionic compound, sodium hydroxide.

APPLICATION

Just as chemists are continually seeking to expand the number of known chemical compounds, nuclear physicists use sophisticated technology to manufacture new elements. Although the 92 elements up to and including uranium are found in nature, the remaining 20 elements have been synthesized by humans. Since the 1940s, physicists have developed special reactors called particle accelerators to induce nuclear reactions that convert one element into another through highly energetic collisions. As implied by its name, particle accelerators use strong electric and magnetic fields to create highly energetic reactant elements that collide at incredibly high velocities. Some of these collisions have been shown to result in the formation of transuranium elements (atomic number >93) not found in nature. In most cases, the transuranium elements formed in particle accelerators are highly unstable and exist for very short time periods (t ≪ 1 second). Therefore, the verification and chemical characterization of newly formed elements can be an extremely difficult endeavor. On the other hand, some transuranium elements are long-lived and can be manufactured in significant quantities. For example, americium-241 is used commercially in smoke detectors. In a sense, physicists have succeeded in the age-old quest of alchemists to convert one element into another element.

DEMONSTRATION PROBLEM

What fraction of the volume of a hydrogen atom (volume = $4.0 \times 10^{-24}\,cm^3$) is taken up by the nucleus if it has a nuclear density of $3.98 \times 10^{14}\,g/cm^3$ and a mass of $1.67 \times 10^{-27}\,kg$?

Solution

Use the density and mass of the hydrogen nucleus to compute volume in units of cm^3. Next, divide the nuclear volume by the total volume of the hydrogen atom to obtain the ratio.

definition of density: density = (mass)/(volume)

rearranging equation: volume = (mass)/(density)

solving for volume: volume = $[(1.67 \times 10^{-27}\,kg) \times (1,000\,g/kg)]/(3.98 \times 10^{14}\,g/cm^3)$
                                   = $4.20 \times 10^{-39}\,cm^3$

calculate ratio: ratio(nucleus/atom) = $(4.20 \times 10^{-39}\,cm^3)/(4.0 \times 10^{-24}\,cm^3)$
                                   = 1.0×10^{-15}

Chapter Test
True/False

1. Electrons possess a negative charge and are located in the nucleus of an atom.

2. Ionic compounds are formed from the association of two or more nonmetal atoms.

3. All the known elements are naturally occurring.

4. Molecular structure gives information pertaining to the order in which atoms are arranged in the form of molecules.

5. $Cr_2O_7^{2-}$ is an example of a polyatomic anion.

Multiple Choice

6. Which one of the following statements about atomic structure is true?
   a. All the mass of an atom comes from its protons.
   b. The number of protons and neutrons are always the same in a neutral atom.
   c. The nucleus of an atom is considerably more dense than the atom itself.
   d. The periodic table organizes the known elements in order of increasing reactivity.
   e. Electrons are positively charged and occupy most of the volume in an atom.

7. Which species below contains a polyatomic cation?
   a. $BaCl_2$
   b. $NaNO_3$
   c. N_2O_5
   d. NH_4Cl
   e. None of the above

8. The charge of the metal cation in $Mn(OH)_2$ is equal to
   a. +1.

b. +2.

c. +3.

d. −1.

e. −2.

9. Which of the atomic symbols below correctly represents an isotope of $^{11}_5B$?

a. ^9_5B

b. $^{11}_6B$

c. $^{11}_4B$

d. $^{11}_5C$

e. $^{11}_4Be$

Short Answer/Essay

10. How many protons, electrons, and neutrons are in the cation $^{24}_{12}Mg^{2+}$?

11. Give the molecular formula for each of the following inorganic compounds: copper (II) bromide, nitric acid, boron trifluoride, and lithium hydroxide.

12. Write the names of the following compounds: SF_6, SnO_2, HCl, and $Ca_3(PO_4)_2$.

Chapter Test Answers

1. **False**

2. **False**

3. **False**

4. **True**

5. **True**

6. **c**   7. **d**   8. **b**   9. **a**

10. 12 protons, 10 electrons, and 12 neutrons

11. $CuBr_2$, HNO_3, BF_3, and LiOH

12. Sulfur hexafluoride, tin(IV) oxide, hydrochloric acid, and calcium phosphate

Check Your Performance

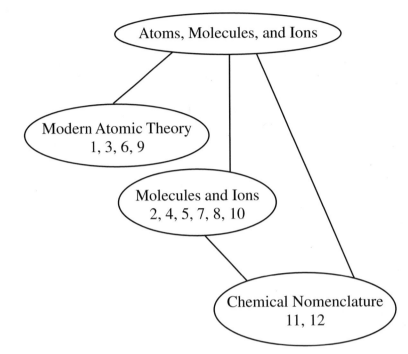

Use this chart to identify weak areas, based on the question numbers you answered incorrectly in the chapter test.

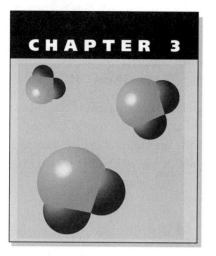

CHAPTER 3

Stoichiometry and Chemical Change

Chemical reactions play an enormous role in all aspects of our daily life. The combustion of fossil fuels heats our homes and powers our cars, chemical reactions in our bodies break food down to sustain life, and even the thought processes associated with reading this page involve complex chemical changes occurring within the synapses of your brain. Over the years chemists have developed a powerful language for describing the important aspects of most chemical reactions. In this chapter, the basics of this language are outlined in terms of chemical stoichiometry. Next, the important consequences of this new concept are reinforced by applying it to a suite of chemical reactions occurring in water solutions.

ESSENTIAL BACKGROUND

- **Dimensional analysis (Chapter 1)**
- **Molecular mass and molecular formula (Chapter 2)**
- **Homogeneous mixtures (Chapter 2)**

TOPIC 1: MOLAR QUANTITY

KEY POINTS

✓ *How is a mole defined?*

✓ *Why are moles used in chemistry calculations?*

✓ *What is Avogadro's number?*

✓ *How does molecular mass relate to molar mass?*

As discussed in the preceding chapter, the materials of our universe can be viewed on a microscopic level as a massive collection of atoms and molecules. Although this microscopic picture of nature is essential in explaining the chemical changes occurring around us, we are often times limited to working with amounts of chemicals that are observable on a macroscopic level or visible by the unaided eye. For this purpose, chemists conveniently work in the special unit of moles. A **mole** is defined as the number of carbon atoms in exactly 12 grams of pure carbon-12. Experimentation over the years has firmly established that this number equals 6.02214×10^{23}, and it is commonly referred to as **Avogadro's number**. The utility of this definition is that it allows us to determine the number of carbon atoms in a sample of ^{12}C by simply knowing its mass. A 42.0-gram ^{12}C sample can be related to the amount of atoms it represents by the following calculation:

$$42.0 \text{ grams} \times (6.02214 \times 10^{23} \text{ atoms}/12 \text{ grams}) = 2.11 \times 10^{24} \text{ atoms}$$
$$\text{Mass} \qquad\qquad \text{Avogadro's Number} \qquad\qquad \text{Number of Atoms}$$

As shown in the calculation above, Avogadro's number can be regarded as a conversion factor $(6.02214 \times 10^{23}\ ^{12}C$ atoms = 12 grams), relating the microscopic world to the macroscopic one. The definition of Avogadro's number can be extended to every element on the periodic table because the masses of two different elements are related to the number of protons and neutrons each possesses. For example, an atom of ^{24}Mg weighs twice as much as one atom of ^{12}C and, thus, 1 mole of ^{24}Mg weighs twice as much as 1 mole of ^{12}C:

$$1 \text{ mole } ^{24}C = 2 \times (1 \text{ mole } ^{12}C) = 2 \times (12 \text{ grams}) = 24 \text{ grams}$$

The fact that the mass of 1 mole of ^{12}Mg is exactly equal to its atomic mass expressed in units of gram is no coincidence. The definition of the mole requires that the mass of 1 mole of any element is equal to the element's atomic mass in units of grams. This mass is called the **molar mass** and is useful as a conversion factor relating the mass of a sample to the number of atoms that comprise it. Because a molecule is merely a collection of atoms, this definition can easily be expanded to derive the molar masses of molecules. A compound's molar mass is simply the molecular mass in units of grams and corresponds to the mass of a sample containing 6.02214 $\times 10^{23}$ particles. For example, the molecular mass of methanol (CH_3OH) is calculated by summing up the masses of the individual atoms of which it is comprised:

$$\text{Molar mass } CH_3OH = (\text{mass of } C) + 4 \times (\text{mass of } H) + (\text{mass of } O)$$
$$= 12 \text{ g/mol} + 4 \times 1 \text{ g/mol} \times 16 \text{ g/mol} = 32 \text{ g/mol}.$$

Topic Test 1: Molar Quantity

True/False

1. The mass of 2.0 moles of methane (CH_4) is 32 grams.

2. The mole is a unit of mass.

3. One mole of glucose ($C_6H_{12}O_6$) contains 6.02214 $\times 10^{23}$ carbon atoms.

Multiple Choice

4. What is the molar mass of KNO_3?
   a. 140.2 amu
   b. 101.1 grams
   c. 101.1 amu
   d. 39.75 grams
   e. None of the above

5. How many atoms of silicon (Si) are in a computer chip of mass 0.354 grams?
   a. 1.56×10^{21} atoms
   b. 2.13×10^{21} atoms
   c. 2.13×10^{-21} atoms
   d. 7.59×10^{21} atoms
   e. 9.912 atoms

6. What is the mass of a sample of radioactive waste containing 1.26×10^{22} uranium-238 atoms?
   a. 0.0209 grams
   b. 209 kilograms
   c. 7.59×10^{45} grams
   d. 7.59×10^{-45} grams
   e. 4.98 grams

Short Answer

7. What is the total number of atoms in 0.156 moles of dimethyl sulfide (CH_3SCH_3)?

8. Why do chemists frequently work in units of moles rather than molecules?

9. Calculate the number of oxygen atoms in a 3.0-pound block of dry ice (solid CO_2) (1 kilogram = 2.2046 pounds).

Topic Test 1: Answers

1. **True.** Molar mass of CH_4 is determined by the sum of the atomic masses in units of grams; molar mass $CH_4 = C + 4H = 12.01\,grams + 4(1.008\,grams) = 16.04\,grams$. Mass of two moles $= 2.0 \times$ molar mass $CH_4 = 2.0 \times 16\,grams = 32\,grams$.

2. **False.** A mole is a unit of amount (how many) not mass (how much).

3. **False.** One mole of glucose ($C_6H_{12}O_6$) contains 6.02214×10^{23} molecules. Because each glucose molecule contains six carbon atoms, the number of carbon atoms can be calculated by the following:

   number of carbon atoms $= (6\,atoms\,C/1$ molecule $C_6H_{12}O_6) \times (6.02214 \times 10^{23}$ molecules) $= 3.61328 \times 10^{24}$ carbon atoms.

4. **b.** The molar mass is calculated by summing the atomic masses of the atoms present. Molar mass $= (mass\,of\,K) + (mass\,of\,N) + 3\,(mass\,of\,O) = 39.10\,grams + 14.00\,grams + 48.00\,grams = 101.10\,grams$.

5. **d.** The number of silicon atoms is determined by converting to moles and using the definition of Avogadro's number: 0.354 grams $\times (1$ mole$/28.085\,grams) \times (6.02214 \times 10^{23}\,atoms/1$ mole$) = 7.59 \times 10^{21}$ atoms.

6. **e.** The mass of the uranium-238 sample can be determined by converting to moles and multiplying by the atomic mass ($238\,g/mol$): 1.26×10^{22} atoms $\times (1$ mole$/6.02214 \times 10^{23}\,atoms) \times (238\,grams/1$ mole$) = 4.98$ grams.

7. $(0.156$ moles $CH_3SCH_3) \times (9\,atoms/1$ mole $CH_3SCH_3) \times (6.02214 \times 10^{23}\,atoms/1$ mole$) = 8.45 \times 10^{23}$ atoms.

8. Chemists use units of moles because it represents a size representative of macroscopic quantities of materials.

9. 3.7×10^{25} O atoms. The number of oxygen atoms in the sample is calculated by using the molecular mass of CO_2 ($44.01\,g/mol$). Remember to include a factor to convert moles of CO_2 to moles of O atom (2 moles O/1 mole CO_2).

$$3.0 \text{ pounds} \times (1 \text{ kilogram}/2.2046 \text{ pounds}) \times (1{,}000 \text{ grams}/1 \text{ kilogram})$$
$$\times (1 \text{ mole } CO_2/44.01 \text{ grams}) = 31 \text{ mole } CO_2$$

$$3.1 \text{ mole } CO_2 \times (2 \text{ moles } O/1 \text{ mole } CO_2) \times (6.02214 \times 10^{23} \text{ atoms}/1 \text{ mole})$$
$$= 3.7 \times 10^{25} \text{ O atoms}$$

TOPIC 2: STOICHIOMETRY AND CHEMICAL REACTIONS

KEY POINTS

✓ *What is a balanced chemical equation?*

✓ *How are mole ratios used to predict the amounts of products formed?*

✓ *What is a limiting reagent?*

✓ *How is the actual yield determined?*

One important consequence of atomic theory is that it describes chemical change as a rearrangement of the atoms that make up reacting materials. This rearrangement occurs by breaking and forming chemical bonds. To keep track of the atoms participating in a chemical reaction, it is useful to express the molecular formulae of all species involved in terms of a balanced **chemical equation**. A chemical equation is a symbolic representation of a chemical reaction that indicates the materials consumed and produced. For example, the combustion of natural gas (methane, CH_4) can be represented by the following chemical equation:

$$CH_{4(g)} + O_{2(g)} \rightarrow CO_{2(g)} + 2H_2O_{(l)}$$
$$\text{REACTANTS} \qquad \text{PRODUCTS}$$

The molecular formulae on the left-hand side of the arrow (CH_4 and O_2) are the starting materials for reaction and are referred to as **reactants** or **reagents**. The molecular formulae on the right-hand side represent species formed via reaction and are called **products**. The physical states of reactant and product molecules are often indicated in parentheses to the right of the molecular formulae. g, l, and s corresponded to gas, liquid, and solid, respectively.

The law of **conservation of mass** provides that the total mass of the reactant molecules must equal the mass of products formed during reaction. As a result, the same number and kinds of atoms must appear on both sides of the balanced chemical equation. To balance a chemical reaction, you must select integer coefficients for each molecule in the chemical equation that make the number of atoms of each element equal on the reactant and product sides. There is no exact sequence of rules to balance chemical equations. The best method is one in which each element is sequentially equalized while taking care to maintain all elements previously balanced. Balancing chemical equations can be viewed as an iterative process (refined over and over) of trial and error. Mastery of this skill only comes with extensive practice (see Topic Test 2, Chapter 2, and Unit I Review).

Balanced chemical equations allow us to quantitatively relate the masses of reactants consumed to the masses of products formed throughout the course of reaction. This process is central to the study of chemistry and is referred to as **stoichiometry**. In stoichiometric calculations, the mass of a reactant or product molecule is related to the mass of all other compounds present in the chemical equation. For example, the reaction of N_2 and O_2 in the engine of your car results in the formation of toxic NO_2 via the following chemical equation: $N_{2(g)} + 2O_{2(g)} \rightarrow 2NO_{2(g)}$. To relate one compound to another, chemists work in units of amount (e.g., one molecule of N_2 reacts with two molecules of O_2 to generate two molecules of NO_2). The easiest way of performing such a calculation uses the ratio of coefficients in the balanced chemical equation as a conversion factor called the **mole ratio**. For example, if we know that 33 grams of N_2 reacts with 75 grams O_2, we can use mole ratios to calculate the mass of NO_2 formed and O_2 reacted upon reaction:

$$33\,\text{grams}\ N_2 \quad \times \quad \frac{1\,\text{mole}\ N_2}{28\,\text{grams}} \quad \times \quad \frac{2\,\text{moles}\ NO_2}{1\,\text{mole}\ N_2} \quad \times \quad \frac{46\,\text{grams}}{1\,\text{mole}\ NO_2} \quad = \quad \textbf{110\,grams}\ NO_2$$

mass of N_2   **molar mass**   **mole ratio**   **molar mass NO_2**   **mass of NO_2**

$$33\,\text{grams}\ N_2 \quad \times \quad \frac{1\,\text{mole}\ N_2}{28\,\text{grams}} \quad \times \quad \frac{2\,\text{moles}\ O_2}{1\,\text{mole}\ N_2} \quad \times \quad \frac{32\,\text{grams}}{1\,\text{mole}\ O_2} \quad = \quad \textbf{75\,grams}\ O_2$$

mass of N_2   **molar mass N_2**   **mole ratio**   **molar mass O_2**   **mass of O_2**

Similarly, we can relate the mass of NO_2 formed to the grams of N_2 reacted using the reciprocal of the mole ratio given above (1 mole N_2/2 moles NO_2)s:

$$110\,\text{grams}\ NO_2 \times \frac{1\,\text{mole}\ NO_2}{46\,\text{grams}} \quad \times \quad \frac{1\,\text{mole}\ N_2}{2\,\text{moles}\ NO_2} \quad \times \quad \frac{28\,\text{grams}}{1\,\text{mole}\ N_2} \quad = \quad \textbf{33\,grams}\ N_2$$

mass of NO_2   **molar mass NO_2**   **mole ratio**   **molar mass N_2**   **mass of N_2**

The amount of product calculated from the mole ratio is referred to as the **theoretical yield**. The flow diagram in **Figure 3.1** summarizes the sequence of calculations involved in typical stoichiometry problems.

Stoichiometry Calculation: A → B

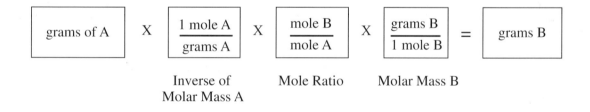

Stoichiometry Calculation with limiting reagent: A + B → C

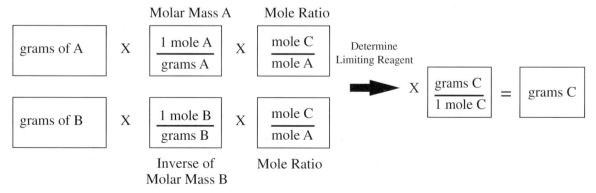

Figure 3.1. Steps in a stoichiometric calculation.

Often, reactants are present in amounts that differ from those defined by the molar proportions in a balanced chemical equation. Under these conditions, only one reactant is expected to be fully consumed at the completion of reaction. Physically, this can be understood as a reaction ceasing because one reactant is fully consumed leaving a portion of the other reactants unused. In this case, the compound that is entirely consumed controls the abundance of products formed and is referred to as the **limiting reactant** or **limiting reagent**. Relating this idea to the previous example, if 33 grams of N_2 is reacted with 33 grams of O_2, we find upon completion that only 14 grams of N_2 have reacted away and 1.0 mole of NO_2 is formed during the reaction.

$$33\,\text{grams } O_2 \times \frac{1 \text{ mole } O_2}{32\,\text{grams}} \times \frac{1 \text{ mole } N_2}{2 \text{ moles } O_2} \times \frac{28\,\text{grams}}{1 \text{ mole } N_2} = \textbf{14\,grams } N_2 \textbf{ reacted}$$

$$33\,\text{grams } O_2 \times \frac{1 \text{ mole } O_2}{32\,\text{grams}} \times \frac{2 \text{ moles } NO_2}{2 \text{ moles } O_2} = \textbf{1.0 mole } O_2 \textbf{ formed}$$

In this example, O_2 acts as the limiting reagent. When performing a stoichiometry calculation in which nonmolar quantities of reactants are present, you must first identify the limiting reagent and then use its amount to calculate the abundance of every other compound in the chemical equation. The flow diagram in Figure 3.1 summarizes the procedures for performing limiting reagent calculations.

Occasionally, the amount of products actually observed upon reaction is less than the amount predicted via stoichiometric calculations. In these cases, the amount observed differs from the theoretical yield and is called the **actual yield**. The ratio of the actual and theoretical yield is often expressed in terms of the percentage yield for a given reaction: **(actual yield)/ (theoretical yield) × 100% = percentage yield**. For example, if we observe the formation of only 0.85 moles of NO_2 under the conditions given in the limiting reagent problem above, we would conclude that the percentage yield of the reaction is 85%: (actual yield)/(theoretical yield) × 100% = (0.85 moles/1.0 mole) × 100% = 85%. Because the actual yield cannot exceed the theoretical yield, the percentage yield will always be less than or equal to 100%.

Topic Test 2: Stoichiometry and Chemical Reactions

True/False

1. All reactant molecules are consumed upon completion of a chemical reaction.

2. The molar mass of a compound represents the mass of 6.02214×10^{23} molecules.

3. The limiting reagent is always the reactant with the smallest mass.

Multiple Choice

4. How many molecules of water are present in a 100.0-gram sample of H_2O?
   a. 5.555 moles
   b. 5.555 molecules
   c. 6.022×10^{25} molecules
   d. 3.346×10^{24} molecules
   e. 9.224×10^{-24} molecules

5. What is the coefficient for H_2O needed to balance the reaction given below?
   $C_6H_{12}O_6 + 6O_2 \rightarrow 6CO_2 + $ ____H_2O
   a. 1
   b. 2
   c. 4
   d. 6
   e. None of the above

Short Answer

6. Ozone (O_3) is a U.S. Environmental Protection Agency-regulated pollutant formed from the reaction of CO and O_2 in most of America's urban centers ($CO + 2O_2 \rightarrow CO_2 + O_3$). Calculate the mass of ozone expected to form upon the reaction of 15.0 grams of CO with an excess of O_2.

7. Balance the following chemical equation used in the industrial production of NH_3:

$$\_\_\_\_\_ N_{2(g)} + \_\_\_\_\_ H_{2(g)} \rightarrow \_\_\_\_\_ NH_{3(g)}$$

8. Ammonium sulfate is formed by the following reaction: $2NH_4^+ + SO_4^{2-} \rightarrow (NH_4)_2SO_4$. How many grams of ammonium sulfate are expected to form when 14.0 moles of NH_4^+ reacts with 6.00 moles of SO_4^{2-}?

Topic Test 2: Answers

1. **False.** In many cases reactants are not present in exact stoichiometric amounts. Thus, one reactant (the limiting reagent) will be consumed and a portion of the others will remain upon completion.

2. **True.** Molar mass of a compound is the mass of 1 mole or 6.02214×10^{23} molecules.

3. **False.** The limiting reagent is the reactant that is entirely consumed upon completion of a chemical reaction. The compound with the smallest mass is not always the limiting reagent.

4. **d.**

$$100.0 \text{ grams} \times \frac{1 \text{ mole}}{18 \text{ grams}} \times \frac{6.02214 \times 10^{23} \text{ molecules}}{1 \text{ mole}} = \textbf{3.346} \times \textbf{10}^{\textbf{24}} \textbf{ molecules of H}_2\textbf{O}$$

5. **d.** A coefficient of 6 makes the number of O atoms (18) and H atoms (12) equal on both sides of the chemical equation.

6. 25.7 grams O_3. The mass of ozone expected to be formed is determined by the following calculation:

$$\text{mass } O_3 = 15.0 \text{ grams CO} \times \frac{1 \text{ mole CO}}{28 \text{ grams}} \times \frac{1 \text{ mole } O_3}{1 \text{ mole CO}} \times \frac{48 \text{ grams } O_3}{1 \text{ mole } O_3}$$

$$= \textbf{25.7 grams O}_3$$

7. An iterative method of trial and error must be used to balance the equation. In the balanced equation:

$$\underline{\quad} \; N_{2(g)} + \underline{\;3\;} \; H_{2(g)} \rightarrow \underline{\;2\;} \; NH_{3(g)}$$

the number of atoms of each type is equal on reactant and product sides of the equation.

8. 793 grams. In this example, SO_4^{2-} is the limiting reactant. This is determined by calculating the amount of NH_4^+ required to react with all the SO_4^{2-} present using the mole fraction: 2 moles NH_4^+/1 mole SO_4^{2-}.

$$6.00 \text{ moles } SO_4^{2-} \times (2 \text{ moles } NH_4^+/1 \text{ mole } SO_4^{2-}) = 12.0 \text{ moles } NH_4^+$$

Because NH_4^+ is present in an amount greater than this, SO_4^{2-} will be the limiting reactant. The mass of product formed is calculated using the amount of SO_4^{2-} present in using the mole fraction: 1 mole $(NH_4)SO_4$/1 mole SO_4^{2-}.

$$6.00 \text{ moles } SO_4^{2-} \times (1 \text{ mole } (NH_4)SO_4/1 \text{ mole } SO_4^{2-}) \times (132 \text{ grams}/1 \text{ mole}) = 793 \text{ grams.}$$

TOPIC 3: CHEMICAL CHANGE IN THE AQUEOUS ENVIRONMENT

KEY POINTS

✓ *How does the aqueous phase differ from other states of matter?*

✓ *What are the differences between electrolytes and nonelectrolytes?*

✓ *What are the three most common types of aqueous phase reactions?*

The presence of water on the Earth's surface differentiates our chemical environment from the other planets in our solar system. Indeed, it is believed that the presence of water played a critical role in the evolution of our biosphere and in the development of life. Because water plays important roles in the world around us, chemists have spent a great deal of time characterizing the nature of chemical reactions that take place in water mixtures. This special class of chemical reactions are classified as **aqueous-phase reactions** because they take place in homogeneous mixtures containing water. Such mixtures are referred to as **aqueous solutions** and always consist of an excess of water and one or more dissolved species. In such mixtures, water is called the **solvent** and the dissolved species are referred to as **solutes**. The concentration of a compound dissolved in water is often expressed in units of **molarity (M)**, which is defined by the expression **molarity = (moles solute)/(liters solution)**. For example, dissolving 6 moles of NaCl into enough water to make 2 liters of solution yields a 3 M solution of NaCl:

$$\text{Molarity} = (\text{moles NaCl})/(\text{liters solution}) = 6 \text{ moles}/2 \text{ liters} = 3 \text{ M}$$

The solution properties of water largely arise from its molecular structure, which is shown in **Figure 3.2**. This geometry indicates that two hydrogen atoms are bonded to oxygen in V-shaped or bent geometry. Although the chemical bonds in water are covalent, the electrons in each bond are preferentially attracted to the oxygen atom. As a result, the oxygen atom in water possesses a slight negative charge, indicated by the symbol δ^-. This attraction also gives rise to slight positive charges on the two hydrogen atoms. As a result of this unequal charge distribution, water is a **polar molecule** with both a positive end and a negative end. Because of its partial charge, water molecules are able to surround certain ionic compounds and disrupt the

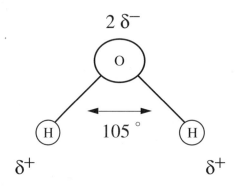

Figure 3.2. Structure and polarity of water.

ionic bonds that hold them together. This process is called hydration or solvation and may be represented by the following chemical equation using NaCl as an example: $NaCl_{(s)} \rightarrow Na^+_{(aq)}$ $+ Cl^-_{(aq)}$. When an ionic compound dissociates to form ions in solution, the resulting mixture is able to conduct electricity. Species that readily produce ions in solution are called **electrolytes** and compounds that produce few or no ions in solution are called **nonelectrolytes**.

Many aqueous-phase chemical reactions are driven by the reactivity of ions formed upon dissolution. In these cases, the chemical equation describing the reaction can be arranged in a number of ways to reflect the reactivity of the ions produced in solution. For example, HNO_3 and $NaOH$ are two strong electrolytes that dissociate completely upon addition to water to produce ions. The **molecular equation** for their interaction may be expressed in terms of the electrolytes themselves:

$$HNO_{3(aq)} + NaOH_{(aq)} \rightarrow H_2O_{(l)} + NaNO_{3(aq)}$$

However, the equation above can also be expressed in terms of the major ionic species present during reaction. This representation is called the **complete ionic equation**. Our example above is represented by the following complete ionic equation:

$$H^+_{(aq)} + NO^-_{3(aq)} + Na^+_{(aq)} + OH^-_{(aq)} \rightarrow H_2O_{(l)} + Na^+_{(aq)} + NO^-_{3(aq)}$$

Na^+ and NO_3^- ions appear on both the product and reactant sides of the equation and are called **spectator ions** because they play no role in the chemical reaction. Omitting the spectator ions yields the **net ionic equation** describing reaction. In our example, the net ionic equation is $H^+_{(aq)} + OH^-_{(aq)} \rightarrow H_2O_{(l)}$.

The reaction between HNO_3 and $NaOH$ is an example of a **strong acid/strong base neutralization reaction**. A strong acid, such as HNO_3, can be thought of as a species that dissolves in water to produce H^+ cations and a strong base such as $NaOH$ is a species that dissolves in water to produce OH^- anions in solution. Weak acids and bases are species that are only partially dissociated in solution. These definitions are refined considerably in Chapter 12. Strong acids and strong bases react to form water via the transfer of a proton. Thus, the net ionic equation for all strong acid/strong base reactions is $H^+_{(aq)} + OH^-_{(aq)} \rightarrow H_2O_{(l)}$.

Anions and cations dissolved in solution can also associate with each other to form solid ionic compounds. Such aqueous-phase chemical reactions are known as **precipitation reactions** because they involve the formation of an insoluble solid called a **precipitate**. An insoluble compound is one that does not dissolve readily in water. For example, if a solution containing $BaCl_2$ is added to a solution of K_2SO_4, the formation of an insoluble white solid is observed. The formation of a precipitate can be understood in terms of the following complete ionic equation:

$$Ba^{2+}_{(aq)} + 2Cl^-_{(aq)} + 2K^+_{(aq)} + SO^{2-}_{4(aq)} \rightarrow BaSO_{4(s)} + 2Cl^-_{(aq)} + 2K^+_{(aq)}$$

Topic 3: Chemical Change in the Aqueous Environment   33

Table 3.1 Solubility Rules for Simple Salts in Water

IONS	RULE	EXCEPTIONS
Cl^-, Br^-, and I^-	Most chloride, bromide, and iodide salts are soluble.	Salts containing Ag^+, Pb^{2+}, and Hg_2^{2+} are insoluble.
NO_3^-	Most nitrate salts are soluble.	
SO_4^{2-}	Most sulfate salts are soluble.	$BaSO_4$, $HgSO_4$, $PbSO_4$, and $CaSO_4$ are insoluble.
OH^-	Most hydroxide salts are slightly soluble.	$NaOH$ and KOH are soluble.
S^{2-}, CO_3^{2-}, CrO_4^{2-}, and PO_4^{2-}	Most salts of these ions are slightly soluble.	
Alkali metal ions (Li^+, Na^+, K^+, etc.)	Most salts of these ions are soluble.	
NH_4^+	Most ammonium salts are soluble.	

This complete ionic equation indicates that Ba^{2+} cations bond readily to SO_4^{2-} to yield a new ionic compound that does not dissolve in water. To predict the complete ionic equation for a precipitation reaction, one must determine the solubility of all possible products of a given chemical reaction. Empirical guidelines help considerably when predicting the solubility of a given ionic compound and are summarized in **Table 3.1**.

In addition to neutralization and precipitation reactions, **oxidation/reduction (or redox)** reactions are readily observed in nature. The driving force in oxidation/reduction reactions is the transfer of electrons from one species to another. The loss of electrons by one species is called **oxidation** and the subsequent gain of electrons by another substance is called **reduction**. This can be easily remembered using the pneumonic OIL RIG, which stands for "oxidation involves loss" and "reduction involves gain." Consider the corrosion of iron by lead (II) sulfate that occurs by the reaction $Fe_{(s)} + PbSO_{4(aq)} \rightarrow FeSO_{4(aq)} + Pb_{(s)}$. By expressing this reaction in terms of its net ionic equation, $Fe_{(s)} + Pb^{2+}_{(aq)} \rightarrow Fe^{2+}_{(aq)} + Pb_{(s)}$, it is easy to see that the driving force in this reaction is the transfer of two electrons from Fe to Pb^{2+}. The net ionic equation in a redox reaction can be viewed as the sum of two **half reactions** that describe oxidation and reduction processes separately:

$$\text{Oxidation half reaction: } Fe_{(s)} \rightarrow Fe^{2+}_{(aq)} + 2e^- \quad (Fe_{(s)} \text{ loses two electrons})$$

$$\text{Reduction half reaction: } Pb^{2+}_{(eq)} + 2e^- \rightarrow Pb_{(s)} \quad (Pb^{2+} \text{ gains two electrons})$$

Oxidation/reduction reactions occurring in aqueous solution are often incorrectly balanced by trial and error methods. To aid in balancing complex redox reactions, it is useful to balance the individual oxidation and reduction half reactions separately and sum them appropriately. The addition of H^+ cations and H_2O is useful when balancing redox half reactions in acidic environments and the addition of OH^- and H_2O is helpful in balancing redox half reactions in basic environments. **Table 3.2** summarizes a basic strategy to be used when balancing oxidation/reduction reactions in neutral, acidic, and basic solutions.

To keep track of the transfer of electrons in redox reactions, it is essential to think in terms of an element's oxidation state. The **oxidation state** reflects the number of electrons associated with an element in a particular chemical form. The oxidation state of an atom in its elemental form is assigned a value of zero. For example, the oxidation state of iron in $Fe_{(s)}$ is 0. Monatomic ions are given an oxidation state equal to the charge they possess. Thus, $Fe^{2+}_{(aq)}$ is assigned an oxidation state of +2. Oxygen, fluorine, and hydrogen in covalent compounds are assigned oxida-

Table 3.2 Steps in Balancing Oxidation/Reduction Reactions
1. Write out individal half reactions describing oxidation and reduction separately.
2. Balance each half reaction.
a. Balance all elements except H and O.
b. Balance O using H_2O.
c. Balance H using H^+.
d. Balance charge using e^-.
3. Equalize the number of electrons transferred by multiplying half reaction by integers.
4. Sum two half reactions. Cancel identical species on product and reactant sides.
5. (For reaction in basic solutions) Add OH^- ions to both sides of the equation equal to the number of H^+ ions present. Form H_2O on the reaction side containing both H^+ and OH^- and eliminate water molecules present on both sides of the equation.

tion states of -2, -1, and $+1$, respectively. Oxidation states for other elements in neutral compounds or polyatomic ions are deduced assuming that the sum of the oxidation states of the elements from which they are composed is either zero for neutral compounds or the species' charge in the case of polyatomic ions. For example, the oxidation state of sulfur in SO_4^{2-} is $+6$:

$$\text{Charge of } SO_4^{2-} = (\text{oxidation state of sulfur}) + 4(\text{oxidation state of O})$$

$-2 = x + 4(-2)$; therefore, $x = +6$.

Topic Test 3: Chemical Change in the Aqueous Environment

True/False

1. All ionic compounds dissolve in water to yield aqueous ions.

2. An acid is a proton donor in a neutralization reaction.

3. All aqueous reactions take place in the presence of water.

Multiple Choice

4. How many moles of ions are formed upon the hydration of 3.50 moles of K_3PO_4 in a water solution? $K_3PO_4 \rightarrow 3K^+_{(aq)} + PO^{3-}_{4(aq)}$
   a. 3.50 moles
   b. 7.00 molecules
   c. 2.10×10^{24} moles
   d. 28.1 moles
   e. 14.0 moles

5. Which species undergoes oxidation in the following redox reaction?
   $CH_4 + 2O_2 \rightarrow CO_2 + 2H_2O$
   a. CH_4
   b. O_2
   c. CO_2
   d. H_2O
   e. The reaction is not an oxidation/reduction reaction.

Short Answer

6. How many grams of NaOH are required to react completely with a 0.101-liter solution of 1.00 M HCl?

7. Give the molecular, complete ionic, and net ionic equations that describe the aqueous-phase reaction of $AgNO_3$ and KBr.

8. Balance the following oxidation reduction reaction in an acidic solution:
   $Cr_2O_7^{2-} + Br^- \rightarrow Cr^{3+} + BrO_3^-$.

Topic Test 3: Answers

1. **False.** Not all ionic compounds are electrolytes. Only compounds soluble in water dissociate into ions in solution.

2. **True.** Strong and weak acids act as proton donors in solution.

3. **True.** An aqueous reaction is one that takes place in a homogeneous mixture containing water.

4. **e.** K_3PO_4 dissociates readily into K^+ and PO_4^{3-} ions in solution. Therefore, 4 moles of ions form from each mole of K_3PO_4. To calculate the moles of ions, 3.50 moles K_3PO_4 × (4 moles ions)/(1 mole K_3PO_4) = 14.0 moles.

5. **a.** The carbon atom in methane goes from an initial oxidation state of −4 to a final oxidation state of +4. This process involves the loss of eight electrons and therefore is oxidation.

6. 4.04 grams. The net ionic equation in a strong acid/strong base neutralization reaction is always $H^+ + OH^- \rightarrow H_2O$. Therefore, the number of grams of NaOH needed can be related to the moles of HCl present by the following calculation:

   Mass NaOH = (0.101 liters HCl) × (1.00 M) × (40.0 grams NaOH)/(1 mole NaOH)
   = 4.04 grams

7. Use Table 3.1 to determine which precipitate forms $(AgBr_{(s)})$.

   Molecular equation: $AgNO_{3(aq)} + KBr_{(aq)} \rightarrow AgBr_{(s)} + KNO_{3(aq)}$
   Complete ionic equation: $Ag^+_{(aq)} + NO^-_{3(aq)} + K^+_{(aq)} + Br^-_{(aq)} \rightarrow AgBr_{(s)} + K^+_{(aq)} + NO^-_{3(aq)}$
   Net ionic equation: $Ag^+_{(aq)} + Br^-_{(aq)} \rightarrow AgBr_{(s)}$

8. $8H^+ + Cr_2O_7^{2-} + Br^- \rightarrow 2Cr^{3+} + BrO_3^- + 4H_2O$.

APPLICATION

Although most acid-base reactions occur in the aqueous phase, some important neutralization reactions occur in the gas phase. For example, several important acids and bases exist as vapors in the Earth's atmosphere. Emissions from automobile exhaust, coal burning, and industrial processes lead to the formation of gaseous nitric acid (HNO_3). In contrast, the principal base in the Earth's atmosphere, ammonia (NH_3), is a gaseous byproduct of agricultural activity. Gaseous HNO_3 and NH_3 react via neutralization to form solid ammonium sulfate: $HNO_{3(g)} + NH_{3(g)} \rightarrow NH_4NO_{3(s)}$. The ammonium sulfate exists as solid particles suspended throughout the Earth's boundary layer, which is the atmospheric region closest to the Earth. The generation of ammonium sulfate particulate has several important environmental consequences. By efficiently scattering solar radiation, ammonium sulfate particulate leads to dramatically reduced visibility in many urban centers in America. Indeed, many of America's cities periodically develop a dense haze due to the formation of ammonium nitrate particulate. Recent epidemiological studies suggest that the presence of urban particulate is also correlated with increasing mortality rates in America's urban centers. Although the medical explanation for the increased mortality rate is unknown, urban centers surrounded by agricultural regions have taken stringent measures to reduce the emissions of the gaseous acids and bases that lead to particulate formation. National regulation measures focusing on controlling the levels of respirable aerosols are also currently under consideration by the U.S. Environmental Protection Agency.

DEMONSTRATION PROBLEM

A 0.03583-mole sample of propanol (C_3H_8O) is combusted with 0.127 moles of O_2 according to the following chemical equation: $2C_3H_8O_{(g)} + 9O_2 \rightarrow 6CO_2 + 8H_2O$. What is the theoretical yield of CO_2 in units of grams?

Solution

First, find the limiting reagent using the molar ratio ($2C_3H_8O/9O_2$). Next, use the appropriate molar ratio to calculate the mass of CO_2 expected upon completion.

Determine limiting reagent:

moles of O_2 required = (0.03583 moles C_2H_8O) × ($9O_2/2C_3H_8O$) = 0.1612 moles O_2

0.1612 moles of O_2 needed < 0.127 moles O_2 given (O_2 is limiting reagent)

Calculate mass of CO_2:

mass CO_2 = (0.127 moles O_2) × ($6CO_2/9O_2$) × (44.01 g/mol) = 3.73 grams CO_2

Chapter Test
True/False

1. Oxidation is the process that involves the gain of electrons.

2. The oxidation state of O in O_2 is zero.

3. A strong acid and strong base will always react to form water in solution.

4. The oxygen atom in a water molecule has a partial positive charge.

5. A 33-gram sample of CO_2 contains 0.75 moles of molecules.

Multiple Choice

6. What is the correct coefficient for $AlCl_3$ in the following chemical equation?
   $$2Al + 6HCl \rightarrow \underline{\quad} AlCl_3 + 3H_2$$
   a. 1
   b. 2
   c. 3
   d. 4
   e. 5

7. How many moles of electrons are transferred in the following oxidation half reaction?
   $$2SO_4 \rightarrow S_2O_8^{2-}$$
   a. 1
   b. 2
   c. 3
   d. 0
   e. 6

8. How many grams of water are produced upon adding 0.103 moles of H_2SO_4 to 0.154 moles of $NaOH$?
   a. 2.77 grams
   b. 1.85 grams
   c. 3.71 grams
   d. 5.54 grams
   e. None of the above

9. Which of the following net ionic equations best describes the reaction produced upon adding $(NH_4)_2SO_4$ to a $HgCl_2$ solution?
   a. $NH_{4(aq)}^+ + Cl_{(aq)}^- \rightarrow NH_4Cl_{(s)}$
   b. $Hg_{(aq)}^{2+} + SO_{4(aq)}^{2-} \rightarrow HgSO_{4(s)}$
   c. $(NH_4)_2SO_{4(aq)} + HgCl_{2(aq)} \rightarrow NH_4Cl_{(s)} + HgSO_{4(s)}$
   d. $(NH_4)_2SO_{4(aq)} + HgCl_{2(aq)} \rightarrow NH_4Cl_{(s)} + HgSO_{4(aq)}$
   e. No reaction occurs

Short Answer/Essay

10. What is the mass of a sample of water consisting of 1,000 H_2O molecules?

11. If 10 milliliters of 0.143 M HCl is added to 100 milliliters of 0.012 M $Ba(OH)_2$, which species acts as the limiting reagent in the resulting neutralization reaction?

12. Balance the following oxidation/reduction equation occurring in a basic medium.
    $$MnO_4^- + Cl^- \rightarrow MNO_2 + ClO_3^-$$

13. What is the concentration (in units of molarity) of a solution made by dissolving 472 grams of $Pb(NO_3)_2$ in 47.2 liters of water?

14. Calculate the mass of $CaSO_4$ formed upon the addition of 501 milliliters of a 1.00 M $CaNO_3$ solution to 301 milliliters of a 2.00 M Na_2SO_4 solution.

Chapter Test Answers

1. **False**
2. **True**
3. **True**
4. **False**
5. **True**
6. **b** 7. **b** 8. **a** 9. **b**
10. 3×10^{-20} grams
11. HCl
12. $2MnO_4^- + Cl^- + H_2O \rightarrow 2MnO_2 + ClO_3^- + 2OH^-.$
13. $3.02 \times 10^{-2} M$
14. 68.2 grams

Check Your Performance

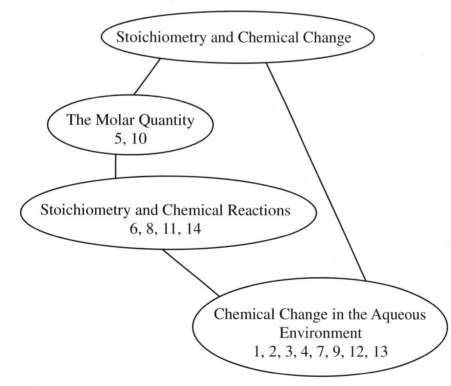

Use this chart to identify weak areas, based on the question numbers you answered incorrectly in the chapter test.

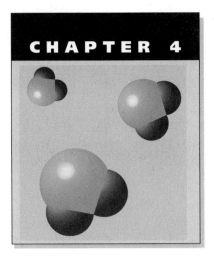

CHAPTER 4

Energy in Chemical Change

Accompanying every chemical reaction is a redistribution of energy stored in the chemical bonds that compose matter. On an annual basis, more than 1,000 million tons of coal, 6,500 million barrels of oil, and 22 trillion cubic feet of natural gas are combusted nationwide to provide energy for heating homes, flying planes, powering factories, and driving cars. Indeed, over 80% of our society's energy is directly harnessed from chemical reactions. The energy change associated with a reaction is not merely a consequence of chemical change but a driving force in determining the spontaneity of chemical processes. Therefore, to understand the motivating forces behind chemical change, we must investigate the subtle role of energy in chemical reactions. In this chapter we discuss chemical change in the context of thermodynamics, which is the study of energy and how it is converted from one form to another.

ESSENTIAL BACKGROUND

- **Dimensional analysis (Chapter 1)**
- **Chemical bonds (Chapter 2)**
- **Balancing chemical equations (Chapter 3)**

TOPIC 1: CHEMICAL NATURE OF ENERGY

KEY POINTS

✓ *What is energy? What units of measure are used in reporting energy?*

✓ *How does potential energy differ from kinetic energy?*

✓ *What is the first law of thermodynamics?*

✓ *What is a state function?*

In the context of chemical reactions, **energy** can be defined as the ability to perform work or produce heat. Therefore, energy is not a material object but rather a property of matter. Energy is typically reported in units of **Joules (J)**, which is defined in terms of fundamental SI units as a kg m^2/sec^2. The Joule is related to the commonly encountered energy unit of calorie by the equivalence statement 1 calorie = 4.184 J. The **first law of thermodynamics** specifies that although energy can be converted from one form to another, it can never be created or destroyed. This law is also commonly called the **conservation of energy**. It is often useful to classify energy as either kinetic energy or potential energy. **Kinetic energy (E_K)** is the energy of a particle associated with its motion and can be calculated using the expression $E_K = 1/2\,mv^2$. **Potential energy (E_P)** is the energy of a particle due to its characteristic position and composition. For example, when methane (CH_4) is burned, the heat produced comes directly from potential energy stored in the C—H bonds.

The total energy of a particle is called **internal energy (E)** and is merely the sum of potential and kinetic energy contributions. It is the change in internal energy accompanying a reaction that is of greatest interest to a chemist. This change can be directly related to the initial and final states ($\Delta E = E_{final} - E_{initial}$) of a chemical system under investigation. The term system in this definition designates the collection of molecules undergoing a particular chemical change. In contrast, the surroundings represent the rest of the matter comprising our universe. Our definition of internal energy is an example of a state function. A **state function** is a property that depends only on its present state and does not depend on the pathway that was used to achieve the present state. The sign of ΔE indicates whether energy has been released ($\Delta E < 0$) or absorbed ($\Delta E > 0$) by the system.

Performing work and exchanging heat are the two most important mechanisms in which energy moves into or out of a chemical system. Therefore, change in internal energy (ΔE) can be expressed in terms of the energy exchange between system and surroundings by the equation **$\Delta E = q + w$**, in which q and w represent the energy exchanges due to heat and work, respectively. In the equation above, energy lost is always assigned a negative value and energy gained is always given a positive value. We focus our description of thermodynamics on the flow of heat that accompanies a chemical change. Exothermic processes are characterized by a flow of heat out of the system (q is negative) and endothermic processes are characterized by a flow of heat into the system (q is positive).

Topic Test 1: Chemical Nature of Energy

True/False

1. The potential energy of the universe is constant.

2. The value of q associated with heating a cup of coffee is negative.

3. Energy stored in chemical bonds is an example of kinetic energy.

Multiple Choice

4. A chemical system does 3.2 J of work on the surroundings while absorbing 11.1 J of heat. The change in internal energy of the system is
   a. 14.3 J.
   b. −14.3 J.
   c. 7.9 J.
   d. −7.9 J.
   e. 0.

5. Which of the following processes is an example of an exothermic process?
   a. Melting an ice cube
   b. Freezing liquid water into an ice cube

c. Heating a cup of water

d. Mixing two identical liquids

e. All of the above

Short Answer

6. Express the first law of thermodynamics both verbally and in terms of a mathematical expression.

7. Explain what a state function is and give an example of a property that can be regarded as a state function.

Topic Test 1: Answers

1. **False.** The total energy of the universe is constant. This represents the sum of contributions from potential energy and kinetic energies.

2. **False.** When heating a cup of coffee, heat flows from the surroundings (the microwave perhaps) into the system (the cup of coffee). Therefore, heating a system will always constitute an endothermic process.

3. **False.** Kinetic energy is the energy associated with movement ($E_K = 1/2\,mv^2$). The energy stored in chemical bonds is an example of potential energy.

4. **c.** The change in internal energy is calculated by summing the exchanges of energy in the forms of work and heat. In this case q is 11.1J (positive because energy is entering the system) and w is −3.2J (negative because doing work represents a loss of energy). Therefore, $\Delta E = q + w = 11.1\text{J} - 3.2\text{J} = 7.9\text{J}$.

5. **b.** Freezing is an example of an *exothermic* process because heat is lost by the system to the surroundings.

6. The first law of thermodynamics states that energy can never be created or destroyed only converted from one form to another. This can be expressed mathematically as $\Delta E_{universe} = 0$ or $\Delta E_{system} = q + w$.

7. A state function is a property that depends only on its present state and does not depend on the pathway that was used to achieve the present state. Altitude is an example of a state function. The altitude of a flying plane only depends on its three-dimensional coordinates. The altitude of a plane does not depend on how the plane got to its present state.

TOPIC 2: CALORIMETRY

KEY POINTS

✓ *What is calorimetry? What does a calorimeter measure?*

✓ *What is specific heat capacity? What units is it expressed in?*

✓ *How does molar heat capacity differ from specific heat capacity?*

By carefully measuring the heat exchanged between a system and its surroundings, a great deal of information can be learned about a particular chemical process. A **calorimeter** is a scientific instrument used to measure the heat exchange that accompanies a physical or chemical process. **Calorimetry** is the science involved with quantifying this heat exchange by monitoring the changes in temperature experienced upon reaction. The **heat capacity** of a substance is equal to the amount of heat required to raise the temperature of that substance one Kelvin unit: $C = q/\Delta T$. In this expression, ΔT is the difference between initial (T_i) and final (T_f) temperatures. The definition of heat capacity can be rearranged to give an expression for the heat required (q) to perform any change in temperature: $q = C\Delta T = C(T_f - T_i)$. Heat capacity is an extensive property, which means it depends on the amount of material present. It is often more useful to define heat capacity in terms of a per mole or per gram quantity. **Molar heat capacity** is the amount of heat required to raise 1 mole of a material 1 K. **Specific heat capacity** is a similar quantity, only expressed in per gram units [$J/(gK)$]. As an example, the amount of heat required to raise the temperature of a 22.0-gram sample of aluminum [$s = 0.900 J/(gK)$] from 298 K to 398 K is determined by the following calculation using specific heat capacity:

$$q = C\Delta T = (\text{specific heat}) \times (\text{mass}) \times (T_f - T_i)$$
$$= 0.900\,J/(gK) \times 22.0g \times (398K - 298K) = 1.98 \times 10^3\,J$$

Topic Test 2: Calorimetry

True/False

1. Molar heat capacity is expressed in units of $(J \times K)/\text{mol}$.

2. A calorimeter can be used to determine the heat exchange accompanying a physical change.

3. Heat capacity is an intensive property.

Multiple Choice

4. How much energy is required to raise the temperature of a 2.576-gram gold coin from 25.0 to 67.0°C [specific heat capacity of gold = $0.151 J/(gK)$]?
   a. 26.0 J
   b. 717 J
   c. −717 J
   d. 16.3 J
   e. None of the above

5. 2.010 J of heat is absorbed by a 0.0122-gram sample of air at 298 K. If the specific heat capacity of air is $1.01 J/(gK)$, what is the resulting temperature of the air sample?
   a. 461 K
   b. 135 K
   c. 229 K
   d. 300 K
   e. 1,000 K

Short Answer

6. How much heat does a 200.0-gram cup of coffee need to lose to change temperature from 60.0 to 25.0°C? Assume coffee has the same heat capacity as liquid water [specific heat = 4.18 J/(g K)].

Topic Test 2: Answers

1. **False.** Molar heat capacity is the heat required to raise the temperature of 1 mole of material 1 Kelvin unit. Therefore, molar heat capacity has units of J/(mol K).

2. **True.** A calorimeter can be used to investigate the heat exchanged during any chemical or physical process.

3. **False.** Heat capacity is an extensive property. For example, we expect that more energy is required to raise the temperature of a bathtub full of water 10°C than is needed to raise the temperature of a small cup of water 10°C.

4. **d.** The heat required can be calculated by rearranging the definition of heat capacity ($C = q/\Delta T$):

$$q = C \times \Delta T = s \times m \times \Delta T = 0.151 \, \text{J}/(\text{gK}) \times 2.576 \, \text{g} \times (67°C - 25°C) = 16.3 \, \text{J}$$

5. **a.** The final temperature of the air sample can be determined by adding the expected temperature change to the initial temperature of the sample.

Calculate the expected temperature change:

$$\Delta T = q/C = q/(s \times m) = 2.010 \, \text{J}/[(1.01 \, \text{J}/(\text{gK})) \times (0.0122 \, \text{g})] = 163 \, \text{K}$$

Determining the final temperature:

$$\Delta T = T_f - T_i, \text{ therefore } T_f = \Delta T + T_i = 163 \, \text{K} + 298 \, \text{K} = 461 \, \text{K}$$

6. The definition of heat capacity ($C = q/\Delta T$) can be rearranged to solve for the heat lost (q) in this problem:

$$q = C \times \Delta T = s \times m \times \Delta T = 4.18 \, \text{J}/(\text{gK}) \times 200 \, \text{g} \times (298 \, \text{K} - 333 \, \text{K}) = -2.93 \times 10^4 \, \text{J}$$

Notice that the answer is negative. This indicates that energy leaves the system and the process is an exothermic one.

TOPIC 3: ENTHALPY OF REACTION

KEY POINTS

✓ *How is enthalpy defined? In what units is it expressed?*

✓ *How is Hess's law used to determine ΔH_{rxn}?*

✓ *What are standard enthalpies of formation?*

Most physical and chemical processes that occur on Earth do so under constant pressure conditions (approximately 760 Torr or 1 atmosphere). Therefore, the heat lost or gained (q_p) accompanying processes at a constant pressure is an especially useful property called **enthalpy (*H*)**. Like internal energy, enthalpy is a state function. As such, the change in enthalpy (ΔH) can be conveniently expressed in terms of initial and final states: $\Delta H = H_{final} - H_{initial} = q_p$. When calculating the enthalpy change of a reaction, the initial state corresponds to the reactant molecules and the final state refers to products. The enthalpy change accompanying a chemical reaction is called the **enthalpy of reaction** and is conveniently expressed in terms of the reactant and product enthalpies: $\Delta H_{rxn} = H_{products} - H_{reactants}$. As an example, consider the highly exothermic oxidation of hydrazine (N_2H_4) by dinitrogen tetroxide (N_2O_4), which is used as a propellant reaction in space craft. The reaction can be expressed in terms of its **thermochemical** equation, which specifies the enthalpy change of reaction:

$$2N_2H_{4(l)} + N_2O_{4(l)} \rightarrow 3N_{2(g)} + 4H_2O_{(g)} \qquad \Delta H° = -2,098\,kJ$$

The superscript (°) designates that the enthalpy change given reflects that all reactant and product molecules are in their **standard states**. Standard states of materials are defined as a pressure of 1 atmosphere for gaseous compounds, a concentration of 1 M for species present in solution, and pure substances for solids and liquids. These definitions are summarized in **Table 4.1**. Notice that the enthalpy of reaction is presented as a per reaction quantity. This allows for calculating the enthalpy change expected when any quantity of reactant is consumed. For example, the heat liberated upon reaction of 72.00 grams of N_2H_4 in an excess of N_2O_4 is determined using stoichiometry:

$$q = 72.00 \text{ grams } N_2H_4 \times (1 \text{ mole } N_2H_4/32.00 \text{ grams}) \times (-2,098\,kJ/2 \text{ moles } N_2H_4)$$
$$= -2,360\,kJ$$

In many cases, a given reaction can be thought of as proceeding through a series of reactions that sum to yield the net conversion of reactants to products. The enthalpy change under such circumstances can be obtained by applying **Hess's law of heat summation**. This law states that the enthalpy change for a reaction that proceeds via a series of individual steps is the sum of the individual enthalpy changes for each reaction. For example, the combustion of CH_4 can occur via the following successive processes:

$$\begin{aligned}
2CH_{4(g)} + 3O_{2(g)} &\rightarrow 2CO_{(g)} + 4H_2O_{(g)} & (\Delta H° = -1,038\,kJ) \\
2CO_{(g)} + O_{2(g)} &\rightarrow 2CO_{2(g)} & (\Delta H° = -566\,kJ) \\
\hline
2CH_{4(g)} + 4O_{2(g)} &\rightarrow 2CO_{2(g)} + 4H_2O_{(g)} & (\Delta H° = -1,604\,kJ)
\end{aligned}$$

Notice that the sum of the two individual reactions yields the net equation and that the total enthalpy change is the sum of the individual reaction enthalpies. The power of Hess's law is that it predicts the enthalpy change for the net process independent of the mechanism chosen for calculation. For example, the enthalpy change calculated above is identical to that experienced by the system if the combustion took place in a single step or via some other mechanism. This is a direct consequence of the fact that enthalpy is a state function. Hess's law can also be under-

Table 4.1 Standard States Defined
Elements
The standard state of an **element** is the **most stable form that the element exists** at a pressure of 1 atmosphere and a temperature of 25°C.
Compounds
Standard state of a **gaseous substance** is equal to a pressure of **1 atmosphere**.
The standard state of a **species in solution** is equal to a concentration of 1 M.
The standard state of a **pure liquid or solid** corresponds to the **pure substance itself**.

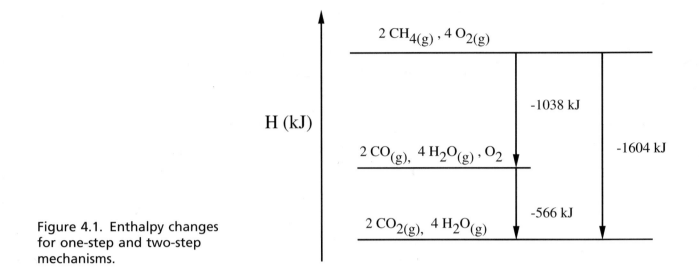

Figure 4.1. Enthalpy changes for one-step and two-step mechanisms.

stood graphically by plotting the initial, final, and intermediate enthalpy states as shown in **Figure 4.1**. There are several important practical considerations when applying Hess's law. The sign of the enthalpy of reaction changes when the corresponding reaction is reversed and the magnitude of the enthalpy of reaction is directly proportional to the molar quantity of reactants consumed.

The standard enthalpy of formation (ΔH_f°) is defined as the enthalpy change that accompanies the formation of 1 mole of a compound in its standard state from its elements in their most stable form in their standard states. For example, the standard enthalpy of formation for CO_2 corresponds to the enthalpy change of the following reaction:

$$C_{(s)} + O_{2(g)} \rightarrow CO_{2(g)} \qquad \Delta H_f^\circ = -394\,kJ/mol$$

The standard enthalpy of formation for elements in their most stable state is arbitrarily scaled to zero, and standard enthalpies of formation are known for most compounds commonly encountered. The application of Hess's law is expanded significantly by using standard enthalpy of formation data. Any chemical reaction can be viewed as a combination of formation reactions that create the reactant and product compounds. For example, the combustion of 2 moles of CH_4 can be written to occur via the following mechanism only involving enthalpy of formation reactions:

$$2CH_{4(g)} \rightarrow 2C_{(s)} + 4H_{2(g)} \qquad (\Delta H_1 = -2\Delta H_f^\circ(CH_4) = 149\,kJ/mol)$$

$$2C_{(s)} + 2O_{2(g)} \rightarrow 2CO_{2(g)} \qquad (\Delta H_2 = 2\Delta H_f^\circ(CO_2) = -787\,kJ/mol)$$

$$4H_{2(g)} + 2O_{2(g)} \rightarrow 4H_2O_{(g)} \qquad (\Delta H_3 = 4\Delta H_f^\circ(H_2O) = -968\,kJ/mol)$$

$$\overline{2CH_{4(s)} + 4O_{2(g)} \rightarrow 2CO_{2(g)} + 4H_2O} \qquad (\Delta H_{rxn} = \Delta H_1 + \Delta H_2 + \Delta H_3 = -1,604\,kJ)$$

Notice that the enthalpy of reaction calculated above is identical to that determined via the former mechanism involving CO. Therein lies the power of Hess's Law. The calculated ΔH_{rxn} is independent of the actual mechanism of reaction. Therefore, the enthalpy change of any reaction can be expressed as the difference between the product standard enthalpies of formation and reactant standard enthalpies of formation. Mathematically, this is represented by the following useful equation: $\boldsymbol{\Delta H_{rxn}^\circ = \Sigma n_p \Delta H_p^\circ - \Sigma n_r \Delta H_r^\circ}$. In this equation, n is the stoichiometric coefficient and the subscripts p and r represent products and reactants, respectively. In combination,

Hess's law and the tabulated standard enthalpies of formation allow chemists to predict the enthalpy changes accompanying almost any conceivable chemical or physical change.

Topic Test 3: Enthalpy of Reaction

True/False

1. Enthalpy is both an extensive property and a state function.

2. Standard enthalpy of formation is independent of physical state.

3. ΔH_f° of O_2 is equal to zero.

Multiple Choice

4. Calculate the ΔH_f° of $CH_3OH_{(g)}$ from the equation below given that ΔH_f° for $CO_{2(g)}$ and $H_2O_{(g)}$ are $-393.5\,kJ/mol$ and $-241.8\,kJ/mol$, respectively.

$$CH_3OH_{(l)} + 2O_{2(g)} \rightarrow CO_{2(g)} + 2H_2O_{(g)} \qquad (\Delta H^\circ = -638.5\,kJ)$$

   a. $-1,274\,kJ/mol$
   b. $-238.6\,kJ/mol$
   c. $-486.8\,kJ/mol$
   d. $486.8\,kJ/mol$
   e. $-974.3\,kJ/mol$

5. Given the thermodynamic equation

$$2CH_{4(g)} + 4O_{2(g)} \rightarrow 2CO_{2(g)} + 4H_2O_{(g)} \qquad (\Delta H^\circ = -1,604\,kJ)$$

   calculate ΔH for the reaction $CO_{2(g)} + 2H_2O_{(g)} \rightarrow CH_{4(g)} + 2O_{2(g)}$
   a. $802\,kJ$
   b. $-802\,kJ$
   c. $1,604\,kJ$
   d. $3,208\,kJ$
   e. $401\,kJ$

Short Answer

6. The body generates energy from the oxidation of glucose ($C_6H_{12}O_{6(s)}$) via the following reaction:

$$C_6H_{12}O_{6(s)} + 6O_{2(g)} \rightarrow 6CO_{2(g)} + 6H_2O_{(l)} \qquad (\Delta H^\circ = -2,803\,kJ)$$

   How much energy can be obtained from the oxidation of 32.0 grams of pure glucose?

Topic Test 3: Answers

1. **True.** Enthalpy is a state function that is proportional to the quantity of reactant consumed. Therefore, it is also an extensive property.

2. **False.** ΔH_f° varies with physical state. For example, ΔH_f° is $-285.8\,kJ/mol$ for liquid water and $-241.8\,kJ/mol$ for gaseous water.

3. **True.** The standard enthalpy of formation is defined as zero for elements in their most stable state.

4. **b.** The ΔH_f° for CH_3OH can be calculated using the relationship $\Delta H_{rxn}^\circ = \Sigma n_p \, \Delta H_p^\circ - \Sigma n_r \, \Delta H_r^\circ$ and the standard enthalpies of formation given.

$$\Delta H_{rxn}^\circ = \Sigma n_p \Delta H_p^\circ - \Sigma n_r \Delta H_r^\circ = \left(\Delta H^\circ(CO_2) + 2\Delta H^\circ(H_2O)\right) - \left(\Delta H^\circ(CH_3OH)\right)$$

rearranging and solving:

$$\Delta H^\circ(CH_3OH) = \left(\Delta H^\circ(CO_2) + 2\Delta H^\circ(H_2O)\right) - \Delta H_{rxn}^\circ$$
$$= (-393.5\,kJ/mol) + 2(-285.8\,kJ/mol) - (-638.5\,kJ/mol)$$
$$= -238.6\,kJ/mol$$

5. **a.** The ΔH° for the reverse reaction can be determined by changing the sign. Next the magnitude must be divided in half because enthalpy is an extensive property (depends on amount). $\Delta H^\circ = -(-1{,}604\,kJ)/2\,mol = 802\,kJ/mol$.

6. $-498\,kJ$. To determine the amount of energy released from reaction, use the enthalpy given as a stoichiometric conversion factor.

$$energy = 32.0\,g \; C_6H_{12}O_6 \times (1 \text{ mole } C_6H_{12}O_6/180\,grams) \times (-2{,}803\,kJ/mol \; C_6H_{12}O_6)$$
$$= -498\,kJ$$

DEMONSTRATION PROBLEM

The process of coal gasification converts solid coal (C_s) to gaseous methane $(CH_{4(g)})$ via reaction with hydrogen gas $(C_{(s)} + 2H_{2(g)} \rightarrow CH_{4(g)})$. The methane produced burns significantly cleaner than coal and is also a more fuel efficient resource. Calculate the heat lost or produced when 1.50×10^7 grams of coal is "gasified" in an excess of H_2 given the following standard heats of formation: $(\Delta H_f^\circ(CH_4) = -74.9\,kJ/mol$ and $\Delta H_f^\circ(C_s) = 0)$.

Solution

First, calculate the enthalpy change for reaction using the standard heats of formation. Remember $\Delta H_f^\circ(H_2)$ is equal to zero because it is the most stable state of elemental hydrogen. Next, convert the mass of coal into moles and use stoichiometry to calculate the energy released to the surroundings upon gasification.

Calculate the ΔH° for reaction:

$$\Delta H^\circ = \Sigma n_p \Delta H_p^\circ - \Sigma n_r \Delta H_r^\circ = \Delta H_f^\circ(CH_4) - \left(\Delta H_f^\circ(2H_2) + \Delta H_f^\circ(C)\right) = -74.9\,kJ$$

Calculate the heat produced for $1.50 \times 10^7 g$ of C:

$$Heat = (1.50 \times 10^7 \text{ grams C}) \times (1 \text{ mole } C/12.01\,grams) \times (-74.9\,kJ/mol \; C)$$
$$= -9.35 \times 10^7 \, kJ$$

Chapter Test
True/False

1. The heat capacities of 1-liter and 10-liter water samples are the same.

2. Potential energy can be converted to kinetic energy.

3. A negative value of q indicates a heat flow into the system.

4. The ΔH_f° of H atom is equal to zero.

5. Molar heat capacity is an extensive property.

Multiple Choice

6. It takes $418\,kJ$ to raise the temperature of 1,000 grams of water from 0.00 to $100°C$. What is the molar heat capacity of water?
   a. $4.18\,J/(K\,g)$
   b. $75,200\,J/(K\,mol)$
   c. $4.18 \times 10^7\,J/(K\,mol)$
   d. $75.2\,J/(K\,mol)$
   e. $89\,J/(K\,mol)$

7. A given chemical reaction does $1.8\,J$ of work on the surroundings while undergoing a change in internal energy (ΔE) equal to $-20.1\,J$. How much heat was transferred to the surroundings during this process?
   a. $18.3\,J$
   b. $20.1\,J$
   c. $1.8\,J$
   d. $21.9\,J$
   e. $0\,J$

8. How many moles of CH_3OH need to be combusted at a constant pressure to generate $112\,kJ$ of heat (i.e., $\Delta H = -112\,kJ$)?

$$2CH_3OH_{(l)} + 3O_{2(g)} \rightarrow 2CO_{2(g)} + 4H_2O_{(g)}\,(\Delta H° = -1{,}455\,kJ)$$

   a. $13.0\,mol$
   b. $0.0769\,mol$
   c. $1.54\,mol$
   d. $0.0770\,mol$
   e. $0.154\,mol$

9. Calculate the $\Delta H°$ for the following reaction given that $\Delta H_f^\circ\,(CO_{(g)}) = -110.5\,kJ/mol$ and $\Delta H_f^\circ(CO_{2(g)}) = -393.5\,kJ/mol$. $2CO_{(g)} + O_{2(g)} \rightarrow 2CO_{2(g)}$.
   a. $-503.5\,kJ$
   b. $-566.0\,kJ$
   c. $503.5\,kJ$
   d. $-283\,kJ$
   e. $283\,kJ$

Short Answer/Essay

10. Calculate the enthalpy of formation for water vapor ($H_2O_{(l)}$) given the following thermochemical equations:

$$H_{2(g)} + 1/2\,O_{2(g)} \rightarrow H_2O_{(g)}\,(\Delta H° = -241.8\,kJ)$$
$$H_2O_{(l)} \rightarrow H_2O_{(g)}\,(\Delta H° = 44.0\,kJ/mol)$$

11. Calculate the heat capacity of a copper (Cu) penny of mass 1.15 grams. [The molar heat capacity of Cu is $24.4\,J/(mol\,K)$.]

12. Which of the following properties are extensive properties? Temperature, heat capacity, enthalpy, internal energy, mass, and concentration.

13. Given the following mechanism, calculate the energy liberated when 44.1 grams of O_3 is converted to O_2.

$$Cl + O_3 \rightarrow ClO + O_2 \quad \Delta H^\circ = -161.5\,kJ$$
$$ClO + O_3 \rightarrow Cl + 2O_2 \quad \Delta H^\circ = -123.8\,kJ$$

Chapter Test Answers

1. **False**

2. **True**

3. **False**

4. **False**

5. **False**

6. **d**   7. **a**   8. **e**   9. **b**

10. $\Delta H^\circ = -285.8\,kJ/mol$

11. $0.442\,J/K$

12. Heat capacity, enthalpy, internal energy, and mass

13. $-131\,kJ$

Check Your Performance

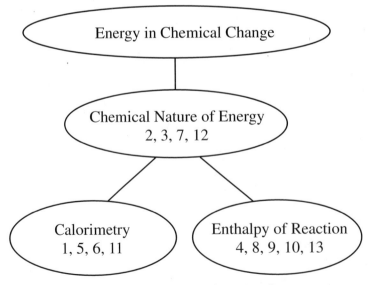

Use this chart to identify weak areas, based on the question numbers you answered incorrectly in the chapter test.

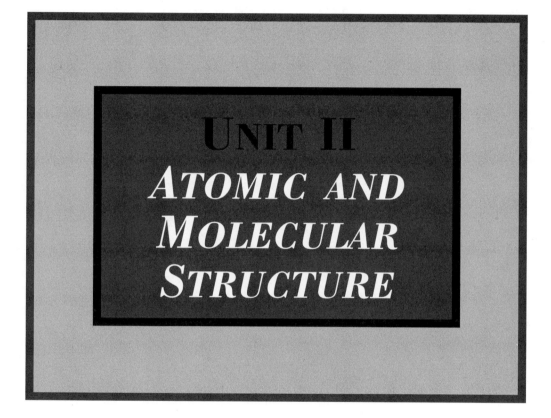

Unit II
Atomic and Molecular Structure

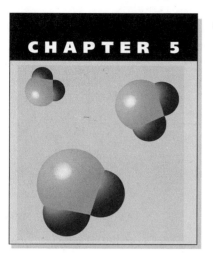

CHAPTER 5

Quantum Description of Matter

Until the early part of the 20th century, scientists assumed that the same physical principles that governed the macroscopic world dictated the actions of atoms and molecules. However, experimentation during the first 20 years of this century cast strong doubt on this notion and established that the microscopic behavior of matter deviated considerably from observations made on the macroscopic level. In this chapter we trace the development of the quantum description of matter by examining the interaction between electromagnetic radiation and atoms. Ultimately, this new description of matter will aid in accounting for similarities and differences in the chemical properties of most elements in the periodic table.

ESSENTIAL BACKGROUND

- **Dimensional analysis (Chapter 1)**
- **Atomic theory (Chapter 2)**
- **The nature of energy (Chapter 4)**

TOPIC 1: PHYSICAL PROPERTIES OF LIGHT

KEY POINTS

✓ *What is electromagnetic radiation?*

✓ *How are wavelength and frequency related?*

✓ *What is a photon? How do its properties differ from waves?*

The development of quantum theory was initiated by a series of revolutionary experiments that used light waves to probe the chemical properties of atoms. Therefore, to understand the foundations of quantum theory, we must first discuss the physical nature of light. Light that is detectable with our eyes is an example of one form of **electromagnetic radiation**. Electromagnetic radiation is a means by which energy travels through space in which it propagates as a wave. Light waves are similar to other types of waves such as ocean waves or sound waves, which consist of oscillations in a physical medium. However, light waves oscillate in electric and magnetic fields that travel throughout space. Light waves consist of repeating patterns of peaks and troughs that are primarily characterized by their wavelength and frequency. **Wavelength** (λ) is the distance between two successive peaks and is measured in units of length. **Frequency** (ν) is

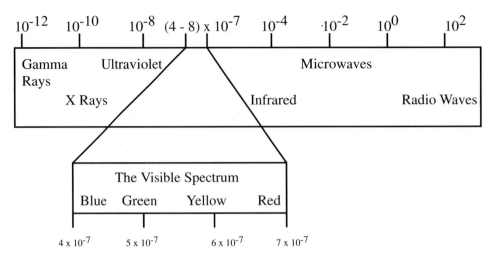

Figure 5.1. Electromagnetic spectrum with wavelength in meters.

the number of wavelength intervals or cycles that pass a given point in space per unit time and is often expressed in units of cycles per second or hertz (Hz). Frequency and wavelength are inversely related because light waves travel at a constant velocity dependent on what medium they propagate through. This relationship is expressed mathematically by the equation $\mathbf{v} = \mathbf{c}/\lambda$, in which c is the speed of light ($c = 3.00 \times 10^8$ m/s in a vacuum). It is useful to characterize light in terms of its position in the electromagnetic spectrum shown in **Figure 5.1**. The electromagnetic spectrum arranges electromagnetic radiation in order of increasing wavelength and is divided up into the following wavelength regions: **gamma rays** (10^{-12} to 10^{-11} m), **x-rays** (10^{-11} to 10^{-8} m), **ultraviolet** (10^{-8} to 10^{-7} m), visible (($4 - 7) \times 10^{-7}$), **infrared** (10^{-6} to 10^{-3} m), **microwaves** (10^{-3} to 10^1 m), and **radio waves** ($>10^1$ m).

An important connection between the frequency of light and its respective energy was established by the theories of two revolutionary physicists, Albert Einstein and Max Planck. By studying the emission of light by materials at high temperatures, Planck concluded that energy radiates from excited materials at specific frequencies dependent on the identity of the emitting material. He established that the frequency of light emitted by a given material follows the expression $\mathbf{E} = \mathbf{nh v}$, in which $\mathbf{v}$ is the frequency of the emitted light wave, h is Planck's constant equal to 6.626×10^{-34} J sec, and n is an integer (1, 2, 3, . . .). Planck's explanation was revolutionary because it suggested that matter was limited to exchange energy in discrete units corresponding to the value of h$\mathbf{v}$. Complimentary to the experiments of Planck, Albert Einstein theorized about a mechanism by which a radiating molecule could produce one unit of light of a given frequency. He proposed that emitted electromagnetic radiation actually consisted of a stream of discrete particles called photons. Further, he hypothesized that each photon had an energy proportional to the frequency of the beam of light. Einstein's explanation suggested that the energy of an individual photon is related to frequency via the expression $\mathbf{E} = \mathbf{h v}$. Einstein provided additional evidence to support the existence of photons by studying the ejection of electrons from illuminated materials that was termed the photoelectric effect. Together, Einstein and Planck had shown that light exhibits properties that are particle-like in addition to their well-recognized wave behavior. This conundrum was central to the development of quantum theory and is called the wave-particle duality of light.

Topic Test 1: Physical Properties of Light

True/False

1. A photon of 750 nm of light has more energy that one with a wavelength of 350 nm.

2. All electromagnetic radiation is detectable with the naked eye.

3. Light exhibits characteristics of particles and waves.

Multiple Choice

4. What is the energy of one photon of red light with a wavelength equal to 725 nm?
   a. 4.81×10^{-40} J
   b. 1.99×10^{-25} J
   c. 2.74×10^{-19} J
   d. 1.99×10^{-34} J
   e. 3.65×10^{18} J

5. A ruby laser emits an intense pulse of 694 nm light. What is the frequency of light produced by a ruby laser?
   a. 4.32×10^{14} Hz
   b. 208 Hz
   c. 4.32×10^{5} Hz
   d. 2.08×10^{20} Hz
   e. None of the above

Short Answer

6. Arrange the following regions of the electromagnetic spectrum in order of increasing energy: visible, radio waves, gamma rays, ultraviolet, infrared, microwaves, and x-rays.

Topic Test 1: Answers

1. **False.** The energy of a photon is inversely proportional to wavelength $E = c/\lambda$. Therefore, a photon of 350 nm light possesses more energy that one of 750 nm.

2. **False.** Visible light makes up a small portion of the entire electromagnetic spectrum with wavelengths from approximately 4.0 to 7.5×10^{-7} m.

3. **True.** Experiments from Einstein and Planck showed that light exhibits both wavelike and particlelike properties.

4. **c.** The energy of a photon of a given wavelength is calculated by combining the equations $E = h\nu$ and $c = \lambda\nu$ to yield the expression $E = (hc)/\lambda$.

$$E = (hc)/\lambda = [(6.63 \times 10^{-34}\, \text{J sec}) \times (3.00 \times 10^{8}\, \text{m/sec})]/[(725\, \text{nm}) \times (1\text{m}/1 \times 10^{9}\, \text{nm})]$$
$$E = 2.74 \times 10^{-19}\, \text{J}$$

5. **a.** Frequency and wavelength are related by the expression $c = \lambda v$. Note: it is essential that units of length cancel correctly in these problems.

$$v = c/\lambda = (3.00 \times 10^8 \, m/sec)/[694 \, nm \times (1m/1 \times 10^9 \, nm)] = 4.32 \times 10^{14} \, Hz$$

6. The order of the electromagnetic spectrum from lowest to highest energy follows the trend of increasing frequency: radio waves, microwaves, infrared, visible, ultraviolet, x-rays, and gamma rays.

TOPIC 2: MODERN QUANTUM THEORY

KEY POINTS

✓ *Why does hydrogen exhibit a line spectrum?*

✓ *What is a wave function? How does it describe the atomic orbitals in atoms?*

✓ *How are uncertainties in a particle's position and momentum related?*

Subsequent experiments probing emission of light from gas mixtures confirmed the theory of quantized energy. Danish scientist Neils Bohr observed that the emission produced upon passing an electric discharge through a sample of hydrogen gas exhibited a discontinuous spectrum composed of specific wavelengths of light. The emission in Bohr's experiment originated from excited hydrogen atoms formed in the discharge and was termed a **line spectrum**. Bohr postulated that each wavelength in the line spectrum resulted from the movement of electrons between allowed energy levels in the hydrogen atom. These energy levels were assigned spherical orbits centered around the nucleus and increased in energy as their distance from the nucleus increased. By characterizing each line present in the H atom emission spectrum, Bohr derived the following expression governing the energy levels available for a single-electron atom:

$$E = -2.178 \times 10^{-18} J \left(\frac{Z^2}{n^2} \right)$$

in which Z is the nuclear charge ($Z = +1$ for H atom) and n is an integer related to the orbital radius. Bohr's energy level equation can be modified to predict the difference in energy between initial ($n_{initial}$) and final (n_{final}) energy levels:

$$E = -2.178 \times 10^{-18} Z^2 \, J \left(\frac{1}{n_{final}^2} - \frac{1}{n_{initial}^2} \right)$$

This energy level model accurately predicts the wavelengths of light emitted from He^+ ($Z = +2$) and Li^{2+} ($Z = +3$) ions in addition to the hydrogen atom.

Although Bohr's theory explained the line spectrum of the hydrogen atom, the theories and experiments of the next generation of physicists eventually replaced it. The resulting model is now called **quantum mechanics** and postulates that the quantization of energy arises from wave-like properties of electrons. Just as light waves can behave as particles, so can matter behave as waves. This radical theory proposes that the wavelength associated with a particle is inversely proportional to its momentum. The wavelength associated with a particle is determined via the expression $\lambda = \mathbf{h}/(m v)$, where m is mass and v is velocity. The wave properties of electrons are quantizied in terms of wave functions. Allowed energy states for an electron are calculated from its wave function using a complex mathematical expression called the Schrodinger wave equation. The wave function also predicts a three-dimensional map of the space occupied by the moving electron called an **atomic orbital**.

The advent and acceptance of quantum mechanics has led to several startling realizations about the microscopic nature of matter. First, the exact location of a moving electron is fundamentally not determinable due to its wave-like properties. This limitation is expressed mathematically in the Heisenberg uncertainty principle: $\Delta x \Delta(mv) = h/(4\pi)$, where Δx and $\Delta(mv)$ are uncertainties in position and momentum, respectively. Second, the location of the electron about the nucleus follows a distribution predicted by probability statistics. The theory of quantum mechanics had for the first time raised scientific arguments that fundamental limitations exist in our ability to perceive the universe on a microscopic level.

Topic Test 2: Modern Quantum Theory

True/False

1. The electron in hydrogen can assume any energy.

2. The observed lines in the hydrogen emission spectrum correspond to specific electronic transitions in the excited atom.

3. The exact location of a moving electron can be calculated using quantum theory.

Multiple Choice

4. What is the wavelength associated with an electron (9.11×10^{-31} kilograms) moving at a speed of 1.8×10^7 m/sec?
   a. 1.1×10^{-70} m
   b. 4.0×10^{-11} m
   c. 2.5×10^{-10} m
   d. 2.5×10^{10} m
   e. 3.7×10^{-18} m

5. What is the difference in energy (ΔE) between the $n = 1$ and $n = 4$ energy levels for the electron in hydrogen?
   a. 2.042×10^{-18} J
   b. 1.623×10^{-18} J
   c. 2.178×10^{-18} J
   d. 1.018×10^{18} J
   e. None of the above

Short Answer

6. Calculate the frequency of light needed to excite an electron in the level $n = 1$ to the level $n = 5$ in the hydrogen atom.

Topic Test 2: Answers

1. **False.** The energy states available to an electron in hydrogen are quantizied. Therefore, they can only assume specific values predicted by the Bohr equation.

2. **True.** The lines in a hydrogen emission spectrum arise from electrons moving from one energy level to another.

3. **False.** One consequence of the wave properties of electrons is our inability to predict both momentum and position simultaneously. Therefore, the position of a moving electron cannot be known absolutely.

4. **b.** The wavelength is determined using the equation: $\lambda = h/(mv)$. The calculation uses the definition of a joule in SI units; $1\,J = 1\,kg\,m^2/sec$,

$$\lambda = (6.626 \times 10^{-34}\,kg\,m^2/sec)/[(9.11 \times 10^{-31}\,kg) \times (1.8 \times 10^7\,m/sec)] = 4.0 \times 10^{-11}\,m$$

5. **a.** The energy difference between any two electronic states in the hydrogen atom ($Z = 1$) is determined using Bohr's energy level equation:

$$\Delta E = E_{final} - E_{initial} = -2.178 \times 10^{-18}\,Z^2\,J(1/4^2 - 1/1^2) = 2.042 \times 10^{-18}\,J$$

6. $3.156 \times 10^{15}\,L/sec$. To calculate the frequency needed, first calculate the energy difference between the $n = 1$ and $n = 5$ states. Next, equate the calculated energy to frequency by the expression $E = hv$.

$$\Delta E = E_{final} - E_{initial} = -2.178 \times 10^{-18}\,Z^2\,J(1/5^2 - 1/1^2) = 2.091 \times 10^{-18}\,J$$

$$E = hv, \text{ therefore } v = E/h = (2.091 \times 10^{-18}\,J) \times (1)^2/(6.626 \times 10^{-34}\,J\,sec) = 3.156 \times 10^{15}\,sec$$

TOPIC 3: ATOMIC ORBITALS

KEY POINTS

✓ *What physical properties are the quantum numbers related to?*

✓ *How do the two electron spin states differ?*

✓ *What does the Pauli exclusion principle state?*

✓ *How are electrons arranged in multielectron elements?*

The atomic orbitals predicted by the Schrodinger wave equation for the hydrogen atom are commonly expressed in terms of a series of numbers called **quantum numbers**. The values of these numbers describe the characteristic energies and three-dimensional shapes of every allowed atomic orbital for a given atom. Each orbital is assigned a unique set of values corresponding to the **principal quantum number** (n: 1, 2, 3, 4 . . . ; determines the orbital energy and distance from the nucleus), **angular momentum quantum number** (l: 0 . . . $n - 1$; defines the orbital shape), and the **magnetic quantum number** (ml: $-l$ . . . l, defines the three-dimensional orientation of the orbital). Values of the angular momentum quantum number (l) are usually expressed in letter equivalents: 0 = s, spherical; 1 = p, dumb-bell shaped; 2 = d, clover shaped; 3 = f, flower shaped. The orbitals available to the electron in hydrogen atom exist in shells determined by the value of n. The energies of these orbitals increase as their distance from the nucleus increases. Within a given shell there exists n^2 unique orbitals that possess the same energy. Orbitals that have the same energy are referred to as **degenerate**. Although degenerate orbitals have the same energy, they may have different shapes and three-dimensional orientations. Typically, orbitals are represented symbolically by giving the value of the principal quantum number followed by the letter representing the angular momentum quantum number.

For example, an orbital described by the quantum numbers $n = 1$ and $l = 1$ is represented by 1s. In addition to the quantum numbers described above, electrons possess a fourth quantum number designating the direction of its electronic spin. The **electronic spin quantum number** (m_s) can possess one of two possible values, +1/2 and −1/2. Although other physical interpretations exist, the values of m_s may be interpreted as indicating two opposite directions in which the electron can spin.

The atomic orbitals in the hydrogen atom can be directly applied to atoms that possess more than one electron. However, in multielectron elements, electron–electron repulsion breaks the degeneracy of orbitals occupying the same shell. As a result of their characteristic electron density distributions about the nucleus, the energies of same shell orbitals in multielectron atoms increase as the angular momentum quantum number is increased. This effect is called electron penetration and gives rise to the following trend in orbital energies for a given shell: $E_s < E_p < E_d < E_f$, where E_x represents the energies of atomic orbitals with different values of the angular momentum quantum number (l).

The arrangement of electrons in multielectron atoms is constrained considerably by the **Pauli exclusion principle**, which states that *no two electrons in the same atom can have the same set of quantum numbers*. This requirement imposes a limitation that each atomic orbital is able to hold a maximum of two electrons corresponding to the two opposite spin states $m_s = 1/2$ and −1/2. The consequences of the Pauli exclusion principle are vast and provide an important key to understanding the arrangement of atoms in the periodic table (see Topic 4).

The most stable arrangement of electrons in an atom is one in which they are located in the lowest energy atomic orbitals available and is referred to as the ground state electron configuration. This arrangement is determined by adding electrons to the atomic orbitals in a multielectron atom in the sequence of lowest to higher energies while obeying the requirements of the Pauli exclusion principle. For example, consider the ground state electron configuration of lithium atom (Li). The three electrons in the ground state exist in the two lowest atomic orbitals, 1s and 2s. This electronic configuration may be symbolically represented by Li: $1s^2 2s^1$. In this representation, the occupied atomic orbitals are written in the order of lowest to highest energy and superscripts indicate the number of electrons in a particular orbital. When more than one configuration exists for a given element, the lowest energy is assigned to the one that has the maximum number of unpaired electrons. In this context, an unpaired electron is one that exists alone in a specific atomic orbital. This condition is called **Hund's rule** and comes into play when placing electrons in degenerate p, d, and f orbitals. For example, the ground state electron configuration of nitrogen (N: $1s^2 2s^2 2p^3$) is shown schematically in **Figure 5.2** and indicates that electrons occupy three 2p orbitals with different orientations. Often, the electronic configuration of an atom is abbreviated by using a noble gas symbol to indicate inner level electrons. An example of this notation for the ground state configuration of N is [He] $2s^2 2p^3$. The electrons in the outermost shell shown in this representation largely determine the atom's chemical properties and are called **valence electrons**.

Topic Test 3: Atomic Orbitals

True/False

1. There are three individual orbitals in the $n = 1$ shell in the hydrogen atom.
2. No two electrons in the same atom can possess the same set of quantum numbers.
3. All orbitals in the same shell are degenerate in a sulfur atom.

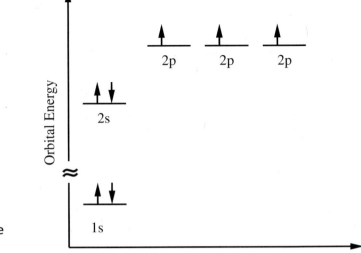

Figure 5.2. Ground state electron configuration of N atom. Arrows designate electrons.

Multiple Choice

4. How many electrons of the same spin can be accommodated in the $n = 2$ shell?
   a. 1
   b. 2
   c. 4
   d. 6
   e. 8

5. Which of the choices below is the ground state electronic configuration of fluorine?
   a. $[He]2s^2 2p^3 d^2$
   b. $[He]2p^7$
   c. $[He]3s^2 3p^5$
   d. $[He]2s^7$
   e. $[He]2s^2 2p^5$

6. How many unpaired electrons are there in the ground state electronic configuration of a sulfur atom?
   a. 0
   b. 1
   c. 2
   d. 4
   e. 5

Short Answer

7. Write the full and abbreviated ground state electron configurations for the following monatomic cations and anions: C^{4-}, Na^+, and Cl^-.

8. How many electrons in a multielectron atom can possess the following sets of quantum numbers?
   $n = 3$
   $n = 3, l = 2$
   $n = 3, l = 2, ml = 1$
   $n = 3, l = 2, ml = 1, m_s = 1/2$

Topic Test 3: Answers

1. **False.** The $n = 1$ shell has 1 orbital (1s) corresponding to quantum numbers $n = 1$, $l = 0$, and $ml = 0$.

2. **True.** This is a restatement of the Pauli exclusion principle.

3. **False.** Sulfur is a multielectron atom. Therefore, the orbitals in a given shell with different values of l have different energies due to electron penetration.

4. **c.** The $n = 2$ shell has a total of n^2 atomic orbitals each with a capacity to hold two electrons: $2 \times (n)^2 = 2 \times 4 = 8$. Half of these electrons may possess the same spin.

5. **c.** Fluorine has a total of seven valence electrons. In the ground state, two of these electrons will be in the 2s orbital and five electrons will occupy the three degenerate 2p orbitals.

6. **c.** All nonvalence electrons in sulfur are paired. Of the six valence electrons, two of these are paired in the 3s and two are paired in the 2p orbital. The remaining two electrons occupy separate 2p orbitals and are unpaired.

7. Electron configurations reflect the number of electrons in each ion rather than the parent element. The noble gas atomic symbol in brackets comprises the abbreviated configurations C^{4-} ($1s^2$ or [He]), Na^{1+} ($1s^2\ 2s^2\ 2p^6$ or [Ne]), and Cl^{1-} ($1s^2\ 2s^2\ 2p^6\ 3s^2\ 3p^6$ or [Ar]).

8. The number of electrons is determined by following the Pauli exclusion principle: *no two electrons in a multielectron atom can have the same set of quantum numbers.*
$n = 3$	: 18 electrons
$n = 3$, $l = 2$	: 10 electrons
$n = 3$, $l = 2$, $m_l = 1$	: 2 electrons
$n = 3$, $l = 2$, $m_l = 1$, $m_s = 1/2$	: 1 electron

TOPIC 4: ELECTRON CONFIGURATION AND THE PERIODIC TABLE

KEY POINTS

✓ *How does the shape of the periodic table reflect electron configuration?*

✓ *What is atomic radius? How does it vary in the periodic table?*

✓ *What is ionization energy? How does it vary in the periodic table?*

✓ *What is electron affinity? How does it vary in the periodic table?*

The ground state electron configuration of an element is determined by successively filling the available atomic orbitals in order of lowest to higher energies. This strategy is typically referred to as the **Aufbau** or "building up" principle and works for almost every element in the periodic table. However, upon reaching the third **period** or row in the periodic table, the relative energies of atomic orbitals in different shells becomes more complex. The energies of atomic orbitals in multielectron atoms have been experimentally shown to obey the following trend from lowest to highest energy: **1s, 2s, 2p, 3s, 3p, 4s, 3d, 4p, 5s, 4d, 5p, 6s, 4f, 5d, 6p, 7s,** and **5f.** This ordering is reflected in the shape of the periodic table in which the position of **groups** is associated with filling different types of atomic orbitals. Groups are columns in the periodic table: s orbitals are designated to the first two groups, p orbitals are associated with the last six groups, and d and f orbitals correspond to the remaining transition elements. These assignments are shown schematically in **Figure 5.3**. Therefore, an element's position in the periodic table directly reflects its electronic configuration and provides an important key to understanding its chemical properties. This characterization is especially useful for predicting three chemical properties: ionization energy, electron affinity, and atomic radius.

Ionization energy is the energy required to remove an electron from an element in its gaseous state. As an example, consider the formation of Na^+ ion via the loss of one electron: $Na_{(g)}([Ne]s^1)$ + 495 kJ/mol $\rightarrow Na^+_{(g)}([Ne])$ + e^-. Removal of the outermost electron is referred to as the first ionization energy and successive electron removals are designated second, third, fourth, and so forth. Experimentation has shown that ionization energy shows a systematic variation with an element's position in the periodic table. Ionization energy increases with atomic number within any given period in the periodic table. This behavior arises because the effective nuclear charge increases as one proceeds from left to right on the periodic table. Effective nuclear charge is the nuclear charge felt by the outer most electrons and its magnitude is directly proportional to the attractive force exerted on these electrons. Ionization energy also exhibits periodic behavior within a given group in the periodic table. It tends to decrease upon proceeding down a given group in the periodic table. This trend reflects the fact that electrons in successive shells are located farther away from the nucleus. Accordingly, such electrons experience weaker attractive forces than electrons closer to the nucleus.

Although atomic orbitals have no abrupt ends, **atomic radius** is a property that describes the size of an atom from a statistical point of view. Atomic radius also shows periodic behavior because it tends to decrease with increasing atomic number within a given period. Similar to the case of ionization energy, this trend is largely due to an increase in effective nuclear charge as one proceeds from left to right within a given row in the periodic table. Atomic radius systematically increases as one moves down a given group in the periodic table because successive periods have additional filled shells of atomic orbitals.

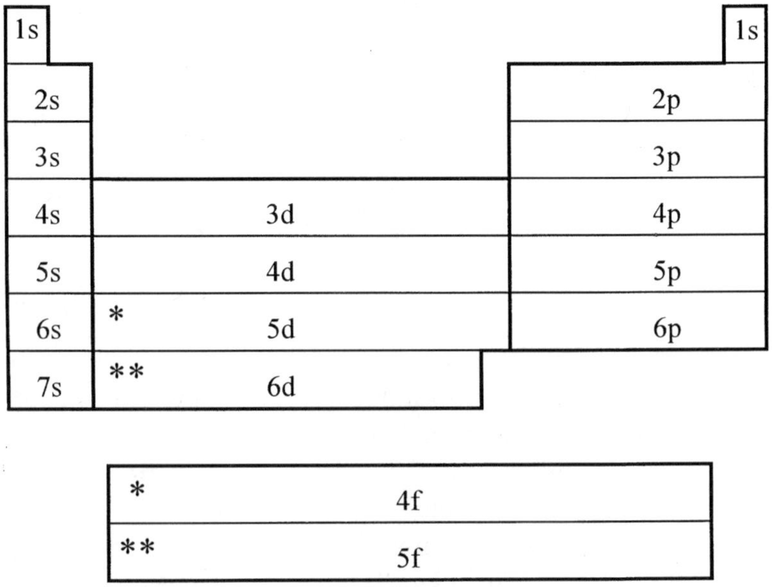

Figure 5.3. Electron configuration and the periodic table.

Electron affinity is the energy change associated with the addition of an electron to an atom in its gaseous state. As an example, consider the formation of F^- ion via the addition of one electron:

$$F_{(g)}([He]2s^2 2p^5) + e^- \rightarrow F^-_{(g)}([He]2s^2 2p^6) \qquad E.A. = -328\,kJ/mol$$

If energy is released during this process, electron affinity is assigned a negative value and if energy is consumed it is designated as positive. The periodic behavior of electron affinity is somewhat more complex than that observed for atomic radius and ionization energy. Although exceptions exist, electron affinities tend toward more negative values, going from left to right across a given period and become more positive proceeding down a group in the periodic table. This behavior can be qualitatively understood in terms of the stability of the anion formed upon addition of an extra electron. Formation of a more stable product anion results in a greater release of energy.

Topic Test 4: Electron Configuration and the Periodic Table

True/False

1. Six of the valence electrons in a ground state iron atom are located in d orbitals.

2. A chlorine atom has a greater atomic radius than a sodium atom.

3. Electron affinity is always a negative quantity.

Multiple Choice

4. Which of the following atoms has the largest ionization energy?
   a. F
   b. O
   c. N
   d. C
   e. B

5. Which of the choices below is the ground state electronic configuration of an iron atom?
   a. $[Ar]\,4s^2\,4p^6$
   b. $[Ar]\,4s^2\,4d^6$
   c. $[Ne]\,4s^2\,4p^6$
   d. $[Ar]\,4s^2\,4p^3\,4d^3$
   e. None of the above

6. Which of the following atoms has the smallest atomic radius?
   a. At
   b. I
   c. Br
   d. Cl
   e. F

Short Answer

7. Write out an equation that describes the chemical process involved with electron affinity using a chlorine atom as an example.

8. Rank the following atoms in order of increasing ionization energy: F, N, C, O, and B.

Topic Test 4: Answers

1. **True.** The electron configuration of ground state Fe is [Ar] $4s^2 4d^6$. This indicates that six electrons are in 4d orbitals.

2. **False.** Atomic radius decreases with increasing atomic number for a given period.

3. **False.** Electron affinity can be a positive or negative quantity depending on whether energy is consumed or released.

4. **a.** The F atom possesses the largest effective nuclear charge. Therefore, it will have the greatest ionization energy.

5. **b.** By using the Aufbau principle, two valence electrons are expected to reside in the 4s orbital and six valence electrons are expected to reside in the 4d orbital.

6. **e.** Atomic radius increases down a group in the periodic table. Therefore, F atom would be expected to have the smallest atomic radius.

7. Electron affinity is the energy released or consumed upon addition of an electron. Thus, the following equation represents the process involved in electron affinity: $Cl + e^- \rightarrow Cl^-$.

8. B < C < N < O < F. This trend follows an increase in the effective nuclear charge proceeding from left to right on the periodic table.

APPLICATION

Lasers are incredibly useful devices that produce an intense beam of light. This light is of a very narrow wavelength range and can be used to send messages in telecommunications systems, read digital information from compact discs (CDs), or accurately measure the passage of time. Typically, chemical lasers consist of an excitation source, material that emits photons and resonance cavity. Depending on how they are designed, lasers operate in pulsed or continuous modes. Visible and ultraviolet light emission results from the transfer of an electron in an exited material to a lower energy state. The electron transition in the excited material is further stimulated by the presence of photons of same frequency as the emitted light. This amplification mechanism results in an intense stream of photons that are in phase with each other or coherent. Because emission occurs under uniform excitation conditions, laser output is usually reproducible and highly stable. Many commercial devices that use lasers take advantage of their inherent stability to decode and transport digital information. For example, CD players use a continuous laser to translate grooves cut onto a CD surface into a digital signal. The laser beam is focused onto the reflective CD surface as it is rotated. When light is reflected from a groove in the CD surface, a decrease in intensity is observed. As the laser beam is tracked across the CD surface, the intensity of the reflected bean is monitored. In this way, the digital signal is decoded. Therefore, stable output of laser light is essential for accurately performing the task of decoding information stored on a CD surface.

DEMONSTRATION PROBLEM

Helium neon lasers are commonly used in commercial devices such as laser pointers. The laser produces a monochromatic beam of photons with a wavelength of 632.8 nm. Calculate the energy of 1.0 mole of photons produced by a helium neon laser.

Solution

To calculate the energy of 1.0 mole of photons of wavelength 632.8 nm, the energy of a single photon of 632.8 nm is multiplied by Avogadro's number.

Calculate the energy of one photon:

$$E = h/c\lambda = (6.626 \times 10^{-34}\,\text{J sec})/[(2.998 \times 10^8\,\text{m/sec}) \times (632.8\,\text{nm}) \times (1\text{m}/1 \times 10^9\,\text{nm})]$$
$$E = 3.493 \times 10^{-36}\,\text{J}$$

Calculate the energy of 1 mole of photons:

$$E = (3.493 \times 10^{-36}\,\text{J}) \times (6.022 \times 10^{23}\,\text{mol}^{-1}) = 2.103 \times 10^{-12}\,\text{J/mol}$$

Chapter Test
True/False

1. Each atomic orbital has the capacity to hold one electron.

2. The three p orbitals in the 2p shell in a nitrogen (N) atom are degenerate.

3. Light possesses both wave-like and particle-like properties.

4. Valence electrons are the unpaired electrons in an atom.

Multiple Choice

5. Which of the electron configurations below indicates the ground state of silicon?
   a. $1s^2\,2s^2\,2p^6\,3s^2\,3p^6$
   b. $1s^2\,2s^2\,2p^6\,3p^4$
   c. $1s^2\,2s^2\,2p^6\,3s^4$
   d. $1s^2\,2s^2\,2p^6\,3s^2\,3p^3$
   e. $1s^2\,2s^2\,2p^6\,3s^2\,3p^2$

6. To what region of the spectrum does lyman alpha emission at 121 nm correspond?
   a. Visible
   b. Ultraviolet
   c. Infrared
   d. Microwave
   e. Gamma ray

7. What wavelength of light is emitted when an Li^{2+} atom undergoes a transition from the $n = 4$ level to the $n = 3$ level?
   a. $2.08 \times 10^{-7}\,\text{m}$
   b. $1.43 \times 10^{15}\,\text{m}$

c. 9.53×10^{-19} m

d. 2.00 nm

e. None of the above

8. How many electrons in a ground state chlorine atom are located in p orbitals?

a. 0

b. 2

c. 5

d. 7

e. 11

9. Which of the following molecules have electrons in s orbitals in their respective ground state electronic configurations?

a. H

b. Cr

c. F

d. Kr

e. All of the above

10. Which of the following atoms has a ground state electronic configuration of [Ar] $4s^2 4d^8$?

a. Pd

b. K

c. Fe

d. Ni

e. None of the above

Short Answer/Essay

11. Write out the electron configurations of the ground states of Mg and Mg^{2+}.

12. Calculate the energy required to remove an electron in the $n = 2$ energy level in the hydrogen atom.

13. Calculate the wavelength associated with a helium nucleus of mass 6.68×10^{-27} kg traveling at a speed of 1.01×10^5 m/sec.

14. The following ground state electron configuration corresponds to which element? $1s^2 2s^2 2p^6 3s^2 3p^6 4s^2 3d^{10} 4p^2$

15. Arrange the following elements in order of increasing atomic radius: P, Br, Ar, Kr.

Chapter Test Answers

1. **False**

2. **True**

3. **True**

4. **False**

5. **e**   6. **b**   7. **a**   8. **e**   9. **e**   10. **d**

11. (Mg)$1s^2 2s^2 2p^6 3s^2$ and (Mg^{2+})$1s^2 2s^2 2p^6$

12. $5.44 \times 10^{-19}\,\text{J}$

13. $9.82 \times 10^{-13}\,\text{m}$

14. Ge

15. Ar < P < Kr < Br

Check Your Performance

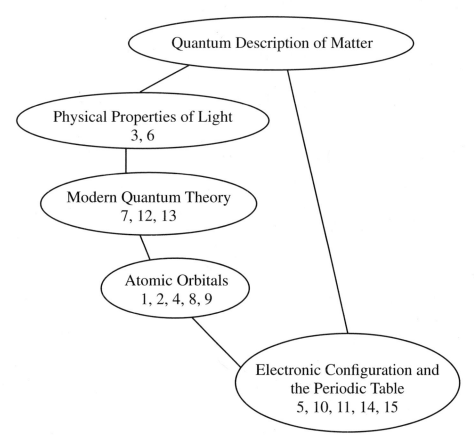

Use this chart to identify weak areas, based on the question numbers you answered incorrectly in the chapter test.

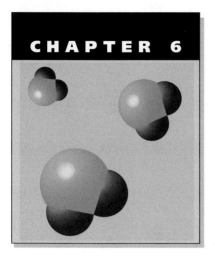

CHAPTER 6

Ionic and Covalent Bonding

The identity, shape, and chemical properties of a molecule are determined by the arrangement of atoms that comprise its structural formula. This three-dimensional arrangement consists of individual atoms bound to each other in a geometry unique to the molecule. Why is molecular structure so important in determining chemical behavior? Accompanying every chemical change is a rearrangement of the atoms that make up the reactant species in the chemical equation. This rearrangement is a complex process involving the simultaneous destruction and formation of chemical bonds. Therefore, the bonds that comprise a particular compound largely determine its chemical reactivity. In this chapter we consider the different types of chemical bonds that hold molecules together to explain their observed chemical and physical behavior.

ESSENTIAL BACKGROUND

- **Atomic theory (Chapter 2)**
- **Cations and anions (Chapter 2)**
- **Enthalpy of reaction and Hess's law (Chapter 4)**
- **Electron configuration and the periodic table (Chapter 5)**

TOPIC 1: IONIC BONDING

KEY POINTS

✓ *What is an ionic bond?*

✓ *How do ions associate to form ionic compounds?*

✓ *What is lattice energy? How can it be understood in terms of Hess's law?*

Chemical bonds are intramolecular forces that hold groups of atoms together in the form of a molecule. Although different types of chemical bonds exist, all bonds arise from the electrostatic interactions of the outermost electrons in the atoms comprising a molecule. An **ionic bond** is a specific type of chemical bond that forms due to the electrostatic attraction between oppositely charged ions. Typically, ionic bonds are formed when one or more valence electrons are transferred from one atom to another atom to form a cation–anion pair. For example, the reaction of sodium metal with chlorine gas to form the ionic compound NaCl occurs via the transfer of a valence electron in Na to Cl: **Na + Cl → Na$^+$ + Cl$^-$ → Na – Cl**. The electronic configurations of all the species involved in the example above are shown in **Figure 6.1**. In this

$$Na \ + \ Cl \longrightarrow \ Na^+ + \ Cl^- \longrightarrow \ NaCl$$

Electron Configurations

[Ne] $3s^1$ [Ne] $3s^2 3p^5$ [Ne]   [Ne] $3s^2 3p^6$

Figure 6.1. Formation of an ionic bond between Na and Cl.

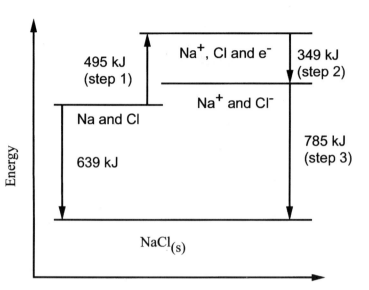

Figure 6.2. Energics involved in the formation of $NaCl_{(s)}$ crystalline lattice.

example, the **ionic compound** formed exists as a uniform **crystalline lattice** in which each Na^+ cation is surrounded by six Cl^- anions. Although the lattice geometry involves coordination of six anions about each cation, the molecular formula simply reflects the overall stoichiometry of the solid. In NaCl, this corresponds to one Na^+ cation for every Cl^- anion. To predict the formula of an ionic compound it is helpful to consider the electronic configurations of the atoms and ions involved in a reaction. The most stable charged state of an ion generally reflects a closed shell electron configuration or noble gas configuration. For example, $Mg([Ne]3s^2)$ and $F([He]2s^2 2p^5)$ atoms are expected to readily form stable $Mg^{2+}([Ne])$ and $F^-([Ne])$ ions that combine to form the ionic compound MgF_2. In this example, two F^- anions are required to bond with one Mg^{2+} to yield a neutral ionic compound.

In the context of the energy associated with bonding, formation of an ionic bond may be viewed as the sum of three separate processes:

1. electron loss to form a cation;

2. electron addition to form an anion;

3. bonding of the cation–anion pair to form a crystalline lattice.

Ionization energy and electron affinity govern the energies associated with cation and anion formation. The energy released upon the formation of an ionic lattice from the reaction of an ion pair in the gas phase is called **lattice energy**. These processes are related to each other via Hess's law and are illustrated graphically in **Figure 6.2**.

Topic Test 1: Ionic Bonding

True/False

1. Electrons are shared between two or more ions in an ionic bond.

2. Lattice energy is the energy released upon the formation of an ionic solid.

3. The most stable ion of oxygen atom is $O^{2-}(1s^2\ 2s^2\ 2p^6)$.

Multiple Choice

4. Which of the molecular formulas below represents the ionic compound formed between Fe^{3+} and SO_4^{2-}?
   a. $FeSO_4$
   b. Fe_2SO_4
   c. $Fe_2(SO_4)_2$
   d. $Fe_2(SO_4)_3$
   e. $Fe_3(SO_4)_2$

5. Which of the following compounds contains an ionic bond?
   a. KBr
   b. CO
   c. H_2
   d. CH_4
   e. All of the above

Short Answer

6. Write out the electron configurations for each of the ions and atoms present in the following reaction: $2Li + O \rightarrow 2Li^+ + O^{2-} \rightarrow Li_2O$.

Topic Test 1: Answers

1. **False.** Ionic bonding involves a complete transfer of electrons from one atom to another. This results in the formation of a mutually attractive cation–anion pair.

2. **True.** Lattice energy is defined as the energy released upon the formation of a crystalline lattice from the reaction of an ion pair in the gas phase.

3. **True.** The electronic configuration of O^{2-} is identical to that of Ne.

4. **d.** To result in a neutral ionic compound, two Fe^{3+} cations must bond with three SO_4^{2-} anions. This is indicated by the subscripts present in the molecular formula $(Fe)_2(SO_3)_3$.

5. **a.** KBr is formed from the association of K^+ and Br^- ions. The other molecules are not held together by ionic bonds.

6. The electron configurations of the stable Li^+ and O^{2-} ions reflect the configurations of noble gases: $\mathbf{2Li([He]2s^1) + O([He]2s^1\ 2p^4) \rightarrow 2Li^+([He]) + O^{2-}([Ne]) \rightarrow Li_2O}$.

TOPIC 2: COVALENT BONDING

KEY POINTS

✓ *What is a covalent bond? How does it differ from an ionic bond?*

✓ *What determines if a covalent bond is polar?*

✓ *How are bond energy and reaction enthalpy related?*

In many substances, electrostatic interactions of valence electrons lead to chemical bonding without the direct transfer of electrons. Indeed, an enormous number of compounds exist such as H_2O, O_2, and CO_2 that are entirely composed of nonmetal atoms in non-ionic states. The attractive forces between atoms in these molecules are called **covalent bonds** and arise from the sharing of electrons between bonded atoms. Covalent bonds form due to the simultaneous electrostatic attraction of electrons to the nuclei of two bonded atoms. For example, consider the covalent bond formed upon the reaction of two hydrogen atoms: $H + H \rightarrow H_2$. As the hydrogen atoms approach each other during reaction, their 1s atomic orbitals overlap in space. Upon overlapping, both electrons are simultaneously attracted to the positive charges of each hydrogen nuclei. These attractive forces hold the atoms in space and constitute the covalent bond in H_2.

The driving force in bonding is the tendency of a chemical system to achieve the lowest energy or most stable state. In terms of the energy of the system, repulsive interactions result in a net increase in energy and attractive interactions results in a net decrease in energy. Accordingly, covalent bonding can be viewed as the sum of three electrostatic interactions that sum to lower the total energy of the system:

1. electron–electron repulsion

2. proton–proton repulsion

3. electron–proton attraction

The length of a covalent bond corresponds to the internuclear distance that minimizes the energy of the system by minimizing electron–electron and proton–proton repulsion while maximizing electron–proton attraction. This distance is called the **bond length** and is the average distance between two bonded nuclei in a molecule. The energy required to break a covalent bond is called the **bond dissociation energy (BDE)** and is approximately equal to the reaction enthalpy (ΔH) of the chemical process comprising bond cleavage. Average values of bond dissociation energies are readily available and are useful for estimating the enthalpy changes for chemical reactions. Under most conditions, the sum of the bond dissociation energies for bonds broken minus the sum of the bond dissociation energies for bonds formed is approximately equal to reaction enthalpy: $\Delta H_{rxn} \cong (\Sigma \text{ BDE (bonds broken)} - \Sigma \text{ BDE (bonds formed)})$. For example, consider the reaction $OH + CH_4 \rightarrow H_2O + CH_3$ that proceeds via a mechanism breaking one C—H (BDE = 435 kJ/mol) bond in methane and forming one OH (BDE = 467 kJ/mol) bond in H_2O. The enthalpy change accompanying reaction can be estimated by the following calculation:

$$\Delta H_{rxn} \cong [\Sigma \text{ BDE (bonds broken)} - \Sigma \text{ BDE (bonds formed)}]$$
$$\cong (435 - 467 \text{ kJ/mol}) \cong -32 \text{ kJ/mol}$$

Although electrons are attracted to the nuclei of both bonded atoms in covalent bonds, the distribution of electron density between them is often not uniform. **Polar covalent bonds** are formed when shared electrons are unequally distributed between two atoms and results in negative and positive poles across the length of a covalent bond. These bonds occur when atoms with different electron attracting properties share electrons. For example, the fluorine atom in HF has a greater ability to attract the shared electrons in its bond with hydrogen. These conditions lead to a partial negative charge on the fluorine atom and a partial positive charge on the hydrogen atom in HF. **Electronegativity** is a measure of the ability of an atom to attract bonded electrons to itself. Although exceptions exist, electronegativity tends to increase as atomic number is increased within a given period and decreases from top to bottom in a given group in the periodic table.

Topic Test 2: Covalent Bonding

True/False

1. The covalent bond in O_2 is polar.

2. Electrons in covalent bonds are attracted to more than one nucleus.

3. Energy is liberated to the surroundings upon bond cleavage.

Multiple Choice

4. Given that the bond dissociation energies of H—Cl and H—O bonds are 427 kJ/mol and 467 kJ/mol, respectively, estimate the enthalpy change for the reaction HCl + OH → Cl + HOH.
   a. 40.0 kJ/mol
   b. −40.0 kJ/mol
   c. 894 kJ/mol
   d. −894 kJ/mol
   e. 0 kJ/mol

5. Which of the following compounds contain a polar covalent bond?
   a. NaCl
   b. N_2
   c. H_2
   d. H_2S
   e. All of the above

Short Answer

6. Given the following thermodynamic equations, calculate the average bond dissociation energy of a C—H bond in CH_4:

$$2H_{2(g)} + C_{(g)} \rightarrow CH_{4(g)} \quad \Delta H = -1,645 \, kJ$$
$$H_{2(g)} \rightarrow 2H_{(g)} \quad \Delta H = 432 \, kJ$$

Topic Test 2: Answers

1. **False.** The covalent bond in O_2 is nonpolar because each O atom has identical electron-withdrawing tendencies or electronegativities.

2. **True.** Electrons are shared between two atoms in covalent bonds due to electron–proton attraction from each nucleus.

3. **False.** Energy is always consumed upon the cleavage of a chemical bond.

4. **b.** The enthalpy change of reaction can be calculated in the following way: ΔH_{rxn} = (Σ BDE (H—Cl) – Σ BDE (H—O) = (427 – 467 kJ/mol) = **–40 kJ/mol**.

5. **d.** H_2S contains two polar covalent H—S bonds.

6. 519 kJ/mol. The reaction involves the cleavage of one H—H bond and the formation of four C—H bonds. Therefore, the average C—H bond dissociation energy is estimated as follows:

Expressing reaction enthalpy in terms of bonds broken and formed:

$$\Delta H_{rxn} = -1,645 \cong [\Sigma \text{ BDE (bonds broken)} - \Sigma \text{ BDE (bonds formed)}]$$
$$\cong [432 \text{ kJ/mol} - 4(\text{BDE C—H})]$$

solving for BDE C—H:

$$\text{BDE C—H} = \textbf{519 kJ}$$

Notice that the sign of bond dissociation energy is always positive.

TOPIC 3: LOCALIZED ELECTRON THEORY

KEY POINTS

✓ *What is a Lewis dot structure?*

✓ *How does the octet rule apply to covalent and ionic bonding?*

✓ *What do resonance structures represent?*

✓ *How is formal charge calculated?*

Localized electron theory is one of the simplest descriptions of chemical bonding. In this theory, shared and nonshared valence electrons are localized in physical regions about the atoms composing a molecule. Electrons localized on a single atom are called lone pairs and electrons located between two atoms are called bonded pairs. The arrangement of lone pairs and bonded pairs of valence electrons are commonly shown in a molecule's **Lewis structure**, also called **Lewis dot structure**. In these diagrams, bonded and nonbonded electron pairs are designated in space by two dots or a dash. Lewis structures of ionic compounds simply indicate the electron configurations of the cation and anion undergoing bonding. Determining Lewis structures of covalently bound molecules involves assigning valence electrons to each atom in a manner that achieves a closed shell or noble gas electron configuration for as many elements as possible. The tendency to achieve a noble gas configuration is often referred to as the **octet rule** because closed valence shells in the second and third periods require a total of eight electrons associated with an atom. In some instances, to achieve this configuration more than two electrons need to be shared by adjacent atoms. Sharing four electrons constitutes a **double bond** and sharing six electrons constitutes a **triple bond**. Exceptions to the octet rule exist and often involve period three and above elements that can attain expanded electronic configurations accommodating more than eight electrons. These expanded electron configurations use d and f orbitals. Although ultimately a trial and error procedure, most Lewis structures can be determined using the following systematic approach:

1. Sum the total number of valence electrons in the molecule.

2. Arrange the elements present in the most plausible orientation and connect with bonded electron pairs.

3. Arrange the remaining electrons around the atoms in bonded and lone pairs to satisfy noble gas electronic configurations for each atom present.

4. When needed, use double and triple bonds between bonded atoms to satisfy this requirement.

Several examples of Lewis structures for ionic and covalent compounds are presented in **Figure 6.3**.

In many cases, more than one equivalent Lewis structure can be drawn for a given molecule. Equivalent structures in localized electron theory have the same number of single and multiple bonds. The resulting series of Lewis structures are called **resonance structures** and indicate

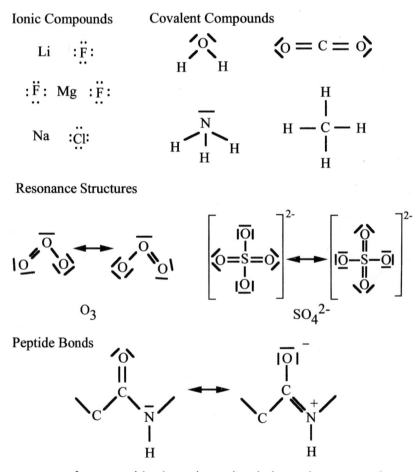

Figure 6.3. Lewis structures for several ionic and covalently bound compounds.

that one or more electron pairs are actually spread over a broader region of the molecule. The resulting electron distribution is referred to as delocalized bonding and is best represented by the average of the resonance structures. Resonance structures are usually written side by side and connected with double-headed arrows. Several examples of delocalized bonding are presented in Figure 6.3.

Occasionally, several nonequivalent Lewis structures exist for a given molecule. In this case, the best Lewis structure may be determined by calculating the charge on each atom in a molecule. **Formal charge** is an estimate of atomic charge that is determined by assigning electrons to each atom indicated in the Lewis structure. Bonded electron pairs are divided equally between adjacent atoms and lone pair electrons are assigned entirely to the atom about which they are localized. Formal charge is calculated by subtracting the number of electrons assigned to the atom from the number of valence electrons in the free atom:

formal charge = (no. of valence electrons in free atom)
$$- [1/2(\text{no. of bonding electrons}) + (\text{no. of lone pair electrons})]$$

An important consequence of this definition is that the sum of formal charges equals zero in a neutral molecule and the charged state in the case of an ion. As an example, consider the resonance structures presented for SO_4^{2-} in Figure 6.3. The formal charge on sulfur is calculated in the following manner:

formal charge = (no. of valence electrons in free atom)
$$- [1/2(\text{no. of bonding electrons}) + (\text{no. of lone pair electrons})]$$
$$= (6 - 4) = +2$$

Formal charge is used to deduce the best Lewis structure by following the guidelines:

1. The best Lewis structure will always reflect the lowest individual formal charges.

2. If a negative formal charge exists, it will preferentially reside on the atom with the greatest electronegativity.

Topic Test 3: Localized Electron Theory

True/False

1. Two bonded pairs and two lone pairs surround the oxygen in H_2O.

2. Localized electron theory can be used to describe ionic and covalent compounds.

3. All molecules have at least two resonance structures.

Multiple Choice

4. What is the formal charge on each oxygen atom bound to sulfur by a single bond in SO_4^{2-} (see Figure 6.3)?
   a. −1
   b. −2
   c. −3
   d. −4
   e. −5

5. Which of the following compounds has a double bond?
    a. NaCl
    b. O_2
    c. H_2
    d. HF
    e. All of the above

6. Which of the following compounds exhibits delocalized bonding (i.e., possesses resonance structures)?
    a. H_2O
    b. NF_3
    c. O_3
    d. HCl
    e. KI

Short Answer

7. Draw Lewis structures for the following compounds: OCS, CF_2S, and C_2H_2.

Topic Test 3: Answers

1. **True.** The oxygen in water is bound to two hydrogen atoms (two bonded pairs) and has two lone pairs (see Figure 6.3).

2. **True.** The Lewis structures of ionic compounds reflect the net transfer of valence electrons and the Lewis structures of covalent compounds indicated sharing of valence electrons.

3. **False.** Only molecules with delocalized electrons possess resonance structures.

4. **a.** Formal charge of each oxygen atom is determined by the following calculation: formal charge = (no. of valence electrons in free atom) − [1/2(no. of bonding electrons) + (no. of lone pair electrons)] = [6 − (1 + 6) = −1].

5. **b.** The two oxygen atoms in O_2 share a total of four electrons and thus are connected by a double bond.

6. **c.** O_3 is the only molecule given that exhibits delocalized bonding. In this case, the resonance structures differ in the position of a double bond.

7. Each atom in the Lewis structures presented below has a closed shell electronic configuration.

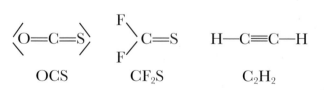

OCS CF_2S C_2H_2

DEMONSTRATION PROBLEM

The combustion of natural gas (CH_4) is an important source of energy in the world and proceeds via the reaction $CH_{4(g)} + 2O_{2(g)} \rightarrow CO_{2(g)} + 2H_2O_{(g)}$. Estimate the enthalpy change accompanying the combustion of methane given the following bond dissociation energies: C—H (411 kJ/mol), O=O (494 kJ/mol), O—H (459 kJ/mol), and C=O (799 kJ/mol).

Solution

To estimate the enthalpy change due to the reaction, one must first determine the Lewis structures of each molecule. Then it is necessary to devise a mechanism proceeding from reactants to products that involves bond breakage and formation. Reaction enthalpy will be approximately equal to the sum of the bond dissociation energies for bonds broken minus the sum of the bond dissociation energies for bonds formed. In the present example, this is determined by the calculation below reflecting a mechanism cleaving four C—H and two O=O bonds and forming two C=O and four O—H bonds.

Estimate the enthalpy change:

$$\Delta H_{rxn} = \Sigma \, BDE \, (\text{bonds broken}) - \Sigma \, BDE \, (\text{bonds formed})$$
$$= \{4(411 \, kJ/mol) + 2(494 \, kJ/mol)\} - \{2(799 \, kJ/mol) + 4(459 \, kJ/mol)\} = \textbf{--802 kJ/mol}$$

APPLICATION

Proteins are macromolecules with masses ranging from 5,000 to 250,000 amu that are entirely composed of smaller subunits called amino acids. They are present in all living cells and perform highly specific cellular functions depending on their composition and chemical environment. The sequence of amino acids that compose proteins is called primary structure and has an enormous bearing on protein's three-dimensional shape and cellular function. In a protein's primary structure, amino acids are connected to each other by strong C—N bonds called peptide bonds resulting in large polypeptide chains. For example, two glycine amino acids react to form a dipeptide via the equation:

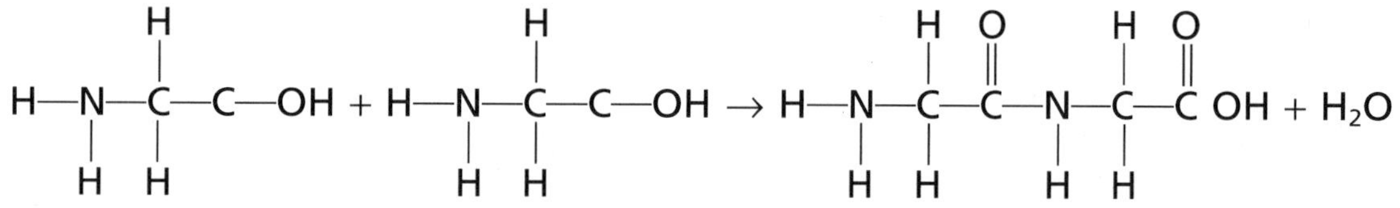

Experiments investigating the strength of polypeptide chains show that the C—N bonds in proteins have unusually large bond dissociation energies. In addition, experimentation has established that rotation around the peptide bonds in polypeptides is hindered, which gives rise to a rigid molecular geometry. One plausible explanation for this behavior is that the peptide bond is better represented by two resonance structures involving alternating double bonds between carbon and oxygen and carbon and nitrogen. These resonance structures are shown in Figure 6.3. Although much is known about the chemical bonding involved in primary structure, the factors that ultimately determine the three-dimensional shape of proteins are extremely complex. Protein structure is in part determined by the interaction of different subunits comprising the polypeptide chain and is currently the subject of intense scientific research.

Chapter Test

True/False

1. Electrons are completely transferred between bonded atoms in a covalent bond.

2. The N in NF_3 is surrounded by three bonded pairs and one lone pair.

3. Energy is consumed upon formation of a covalent bond.

4. An ionic lattice consists of anions and cations held in place by ionic bonds.

5. ΔH is negative for the following reaction: $CH_3CH_2CH_2CH_3 \rightarrow CH_3CH_2CH_2CH_2 + H.$

Multiple Choice

6. Which of the following molecules contains one or more lone pairs?
    a. NaF
    b. H_2
    c. CO_2
    d. CH_4
    e. None of the above

7. Br atoms are most likely to form an ionic bond with which of the following atoms?
    a. Ca
    b. O
    c. Br
    d. Ar
    e. C

8. Which of the following molecules possesses a triple bond?
    a. O_2
    b. CF_4
    c. Br_2
    d. N_2
    e. None of the above

9. Which of the following bonds is the most polar?
    a. H—F
    b. H—Cl
    c. H—Br
    d. H—H
    e. H—I

10. How many lone pairs are present in the Lewis structure for CS_2?
    a. 0
    b. 1
    c. 2
    d. 3
    e. 4

11. How many of the molecules below contain ionic bonds?
    H_2 H_2O CO NH_4 H_2S

a. 0
b. 1
c. 2
d. 3
e. 4

Short Answer/Essay

12. Estimate the enthalpy change for the reaction $2F_2 + 2H \rightarrow 2HF + 2F$, given the bond dissociation energies of F—F (155 kJ/mol) and H—F (565 kJ/mol).

13. How many resonance structures exist for the NO_3^- ion?

14. ICl_4^- is an ion whose Lewis structure does not lead to a noble gas electron configuration for iodine. If the best Lewis structure for ICl_4^- is one in which I shares one bonded pair with each Cl atom and possesses two lone pairs, what is the formal charge of I in this molecule?

15. How many lone pairs are present in the Lewis structure of formaldehyde (CH_2O)?

16. Rank the following bonds in order of increasing bond polarity: H—Br, H—F, H—Cl, H—I.

17. Which of the following molecules contain at least one multiple bond (double or triple bond): N_2, CF_4, SO_2, HCl, NF_3, Br_2, and CH_3OH?

18. Draw Lewis structures for the following compounds: $CaBr_2$, PF_3, and CS_2.

Chapter Test Answers

1. **False**

2. **True**

3. **False**

4. **True**

5. **False**

6. **c**   7. **a**   8. **d**   9. **a**   10. **e**   11. **a**

12. −820.0 kJ/mol

13. 3

14. −1

15. 2

16. H—I < H—Br < H—Cl < H—F

17. N_2 and SO_2

18. :B̈r: Ca :B̈r: |F̄—P—F̄| S̈=C=S̈
                          |
                         |F|

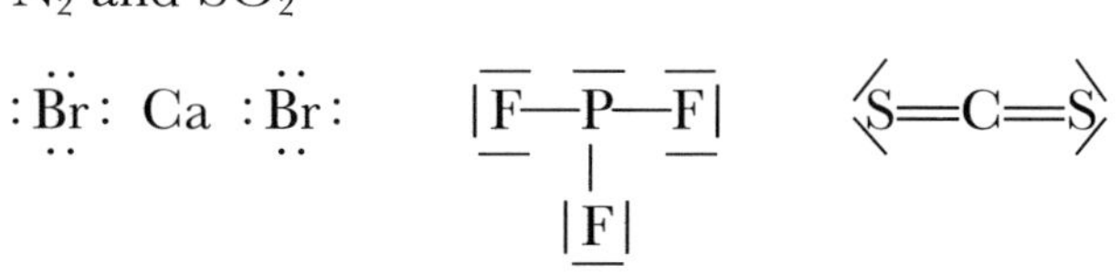

    $CaBr_2$ PF_3 CS_2

Check Your Performance

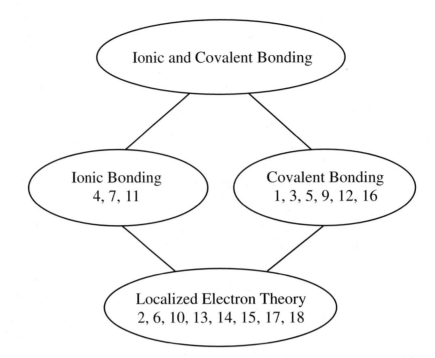

Use this chart to identify weak areas, based on the question numbers you answered incorrectly in the chapter test.

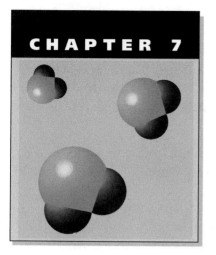

Molecular Geometry and Bonding Theory

CHAPTER 7

The microscopic world is not nearly as static as it may appear from observations made on the macroscopic level. Most molecules are continually vibrating, rotating, and colliding with each other on a very fast timescale. The dynamic nature of the microscopic world is also reflected in a molecule's three-dimensional geometry. Collisions, vibrations, and rotations continually distort the equilibrium positions of atoms comprising a particular molecule. However, equilibrium geometry does have a great bearing on the chemical and physical properties of most molecules. In this chapter, the various factors that give rise to molecular shape are discussed. This description focuses on two important theories describing covalent bonding and molecular geometry: valence shell electron pair repulsion theory and molecular orbital theory.

ESSENTIAL BACKGROUND

- **Atomic orbitals (Chapter 5)**
- **Electronic configuration (Chapter 5)**
- **Localized electron theory (Chapter 6)**

TOPIC 1: VALENCE SHELL ELECTRON PAIR REPULSION THEORY

KEY POINTS

✓ *How is equilibrium molecular geometry predicted?*

✓ *What is a dipole moment?*

✓ *How is the polarity of a molecule predicted?*

Molecular geometry is the size and shape of a molecule in its equilibrium state. Typically, molecular geometry is defined in terms of the angles between adjacent atoms called bond angles and bond lengths that comprise a given molecule. Although molecular geometry can be accurately determined experimentally, several theories exist for predicting the shapes and sizes of molecules. The **valence shell electron pair repulsion (VSEPR) model** is one of the most useful of these theories, in which molecular geometry is determined by minimizing the repulsion of electron pairs localized around the various atoms comprising a molecule. The state of minimum electron repulsion reflects a geometry in which electron pairs reside as far away from each other as possible. Accordingly, the number of lone pairs and bonded pairs about a central atom characterizes a compound's **electron pair geometry**. However, for the purpose of determining electron-pair geometry, double and triple bonds are each counted as one electron pair. For example, the two electron pairs, each a double bond, in $\langle O{=}C{=}O \rangle$ give rise to a **linear** electron pair geometry in which the angles between both bonded pairs are 180 degrees. Other electron pair arrangements are defined by three (**trigonal planar**), four (**tetrahedral**), five (**trigonal bipyramidal**), and six (**octahedral**) electron pairs. **Table 7.1** summarizes the electron pair geometries encountered in many molecules.

The relative position of lone pair electrons and bonded pairs in a compound determines its overall **molecular geometry**. While electron geometry defines the position of electron pairs, molecular geometry refers to the arrangement of atoms in a molecule. Therefore, molecular geometry is characterized by the relative position of bonded electron pairs only. For example, the Lewis structure of H_2O is characterized by an oxygen atom surrounded by two bonded pairs and two lone pairs. The overall molecular geometry appears "bent" because two electron pairs in the tetrahedral arrangement are nonbonding lone pairs. The electron pair geometries and molecular shapes for molecules with up to four electron pairs about a central atom are summarized in **Table 7.2**. Most molecular geometries can be predicted by first drawing a compound's Lewis structure and then determining the electron pair geometries about a central atom:

Lewis structure → electron pair geometry → molecular geometry.

Experimental determinations of the bond angles in many simple molecules deviate slightly from those predicted using the VSEPR model. For example, the bond angle observed in H_2O is smaller than that defined by a tetrahedral electron geometry (104.5 vs. 109.5 degrees). Deviations from the bond angles predicted from VSEPR theory primarily result from lone pairs and multiple bonds, which require more space than single-bonded electron pairs. Lone pair electrons occupy more diffuse spatial regions about an atom than bonded electrons because they are attracted to only one nucleus. Although multiple bonds only count as one repulsive electron pair

Table 7.1 Electron Pair Geometries in Most Simple Molecules

Number of Electron Pairs	Geometry	Bond Angles (degrees)	Examples
2	Linear	180	CO_2, BeF_2
3	Trigonal planar	120	BF_3, O_3
4	Tetrahedral	109.5	H_2O, CCl_4
5	Trigonal bypyramidal	90 and 120	PF_5, SF_4
6	Octahedral	90	SF_6, $XeCl_4$

Table 7.2 Electron Pair and Molecular Geometries for Simple Molecules with a Central Atom Surrounded by up to Four Electron Pairs

Electron Pairs					
Total	Bonding	Nonbonding	Electron Pair Geometry	Molecular Geometry	Example
2	2	0	Linear	Linear	CO_2
3	3	0	Trigonal planar	Trigonal planar	BF_3
	2	1		Bent/angular	SO_2
4	4	0		Tetrahedral	CF_4
	3	1	Tetrahedral	Trigonal pyramidal	NF_3
	2	2		Bent/angular	H_2O

for the determination of geometry, they require more space simply due to the greater number of electrons present. As a result, lone pairs and multiple bonds tend to constrict the arrangement of bonded pairs in their vicinity.

The charge distribution in a molecule is determined by the polarity of its bonds in addition to its overall molecular geometry. **Polar molecules** possess regions of partial positive and negative charges. Molecules that exhibit this property are said to possess a **dipole moment** (μ) that is a vector quantity corresponding to the magnitude and direction of charge separation. The direction of polarity is indicated by the following symbols using the molecule HBr as an example:

$$+ \quad > \qquad \text{or} \qquad \delta^+ \quad \delta^-$$
$$\text{H—Br} \qquad\qquad \text{H—Br}$$

In a molecule possessing three or more atoms, polarity is determined by summing all individual bond dipoles with respect to their orientation (vector summation). Therefore, a symmetric molecule like CF_4 is not a polar molecule despite the fact that each C—F bond exhibits positive and negative poles. In contrast, H_2O is a polar molecule because its O—H bonds do not directly oppose each other. The molecular geometry of water gives rise to a negative pole on the central oxygen atom and a positive pole located exactly between the two hydrogen atoms.

Topic Test 1: Valence Shell Electron Pair Repulsion Theory

True/False

1. C_2H_2 possesses a dipole moment.

2. Lone pairs tend to occupy more space than bonded electron pairs.

3. NH_3 exhibits a trigonal planar molecular geometry.

Multiple Choice

4. Which of the following molecules is nonpolar?
    a. NF_3
    b. H_2O
    c. CO
    d. CO_2
    e. CH_3I

5. What is the electron pair geometry around the sulfur atom in H_2S?
    a. Linear
    b. Bent
    c. Tetrahedral
    d. Trigonal planar
    e. Octahedral

Short Answer

6. Order the following molecules in increasing H—X—H bond angle where X represents N, O, or C atoms: NH_3, H_2O, and CH_4.

7. Give electron pair and molecular geometries about the central atom of the following simple compounds: CF_2O, SO_2, CBr_4, and Cl_2O.

Topic Test 1: Answers

1. **False.** The C—H bonds in C_2H_2 have identical bond polarities but are oriented in opposing directions.

2. **True.** Lone pair electrons are attracted to only one nucleus and thus occupy a more diffuse spatial region about an atom.

3. **False.** The nitrogen atom in NH_3 is surrounded by four electron pairs (three bonded and one lone pair). Therefore, it has a trigonal pyramidal molecular geometry.

4. **d.** The C=O bonds in CO_2 exactly oppose each other, which result in a nonpolar distribution of charge.

5. **c.** There are two bonding pairs and two lone pairs about the sulfur atom. Therefore, H_2S has a tetrahedral electron geometry and a bent molecular geometry.

6. Lone pairs take up more room about a nucleus than bonded pairs. Therefore, ordering from smallest to largest bond angle would be $H_2O < NH_3 < CH_4$.

7. Write out Lewis structures and use Tables 7.1 and 7.2 to deduce electron pair and molecular geometries about the central atom.

   CF_2O: Electron pair (trigonal planar)   **Molecular (trigonal planar)**
   SO_2: Electron pair (trigonal planar)   **Molecular (bent)**
   CBr_4: Electron pair (tetrahedral)   **Molecular (tetrahedral)**
   Cl_2O: Electron pair (tetrahedral)   **Molecular (bent)**

TOPIC 2: HYBRIDIZATION AND BONDING

KEY POINTS

✓ *Which atomic orbitals are used in hybridization?*

✓ *How is the hybridization of a bonded atom predicted?*

✓ *How does hybridization explain expanded octet electron configurations?*

In the VSEPR model, bonding results from the overlap of two atomic orbitals containing electrons of opposite spin. Orbital overlap creates a region of enhanced electron density between atoms that holds the atoms together in space. For most polyatomic molecules, experimentation has shown that the orbitals of bonded atoms differ significantly from the atomic orbitals present in isolated elements. As an example, consider the molecule CF_4. Experimentation has shown that the bonded pairs in this molecule reside in four equivalent orbitals with a tetrahedral geometry. However, the available valence shell atomic orbitals in carbon, 2s, $2p_x$, $2p_y$, and $2p_z$, do not reflect the observed tetrahedral shape. To resolve this discrepancy, the orbitals in bonded atoms are usually thought of as **hybrid orbitals** formed by mixing various atomic orbitals for a particular element. In the CF_4 example, one s and three p atomic orbitals mix to form four orbitals called sp^3 orbitals that surround the central carbon atom in CF_4. The resulting sp^3 bonding orbitals are degenerate and possess the expected tetrahedral geometry. In this example, the hybridization of the central carbon atom can be described schematically by the diagram

$$\uparrow \qquad \uparrow \uparrow \uparrow \text{ (hybridize)} \rightarrow \uparrow \uparrow \uparrow \uparrow$$
$$\text{2s} \quad \text{2p} \qquad\qquad\qquad \text{sp3}$$

It is important to note that the number of atomic orbitals that undergo hybridization always equals the number of hybrid orbitals formed. Hybridization also affects the energies of the bonded orbitals. As shown in **Figure 7.1**, the hybrid sp^3 orbitals have energies between those of the s and p atomic orbitals from which they are derived.

All electron arrangements and molecular geometries predicted by VSEPR theory can be viewed in terms of hybridized atomic orbitals. Bonded and nonbonded electron pairs in VSEPR theory occupy degenerate orbitals formed by hybridizing s, p, and d atomic orbitals. Hybridization of one s and one p orbital forms two **sp orbitals** oriented in linear electron geometry. One s and two p orbitals hybridize to form three degenerate **sp²** orbitals oriented in a trigonal planar electron geometry. Four degenerate **sp³** orbitals are formed by mixing one s and three p orbitals and reflect a tetrahedral geometry. Expanded octet electron configurations needed for trigonal bipyramidal (**sp³d**) and octahedral (**sp³d²**) geometries can similarly be configured using the d orbitals for period 3 and above elements. Although many different hybridization states exist for a bonded atom, the number of atomic orbitals undergoing hybridization must always equal the number of hybridized orbitals formed. The various hybrid orbitals used in explaining VSEPR geometries are summarized in **Table 7.3**.

Most covalent bonds form by the overlap of two hybrid orbitals that concentrate electron density about the internuclear axis. This axis is the line connecting bonded nuclei. These bonds give rise to molecular shape and are called **sigma (σ) bonds**. Sigma bonds comprise almost every single

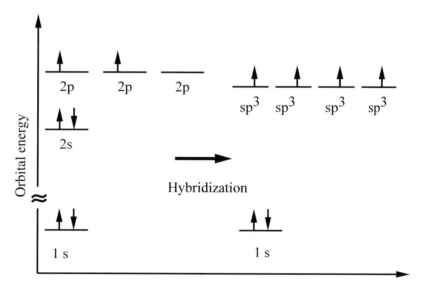

Figure 7.1. Hybridization of the central carbon atom in CF_4.

Table 7.3 Hybrid Orbitals in VSEPR Theory		
STARTING ATOMIC ORBITALS	**HYBRID ORBITALS**	**GEOMETRY**
s + p	sp + sp	Linear
s + p + p	$sp^2 + sp^2 + sp^2$	Trigonal planar
s + p + p + p	$sp^3 + sp^3 + sp^3 + sp^3$	Tetrahedral
s + p + p + p + d	$sp^3d + sp^3d + sp^3d + sp^3d + sp^3d$	Trigonal bypyramidal
s + p + p + p + d + d	$sp^3d^2 + sp^3d^2 + sp^3d^2 + sp^3d^2 + sp^3d^2 + sp^3d^2$	Octahedral

covalent bond encountered in nature. To describe molecules containing one or more multiple bonds, we need to consider a second type of bonding called **pi (π) bonding**. A double bond contains one sigma and one pi bond and a triple bond contains one sigma and two pi bonds. Pi bonds result from the sideways overlap of two unhybridized p orbitals and concentrate electron density above and below the internuclear axis. For example, the triple bond in N_2 can be explained by the overlap of two hybrid sp orbitals to form a sigma bond and the overlap of two pairs of unhybridized p orbitals resulting in two pi bonds. The pi bonds reside above and below the internuclear axis in N_2. The hybridization of each N atom can be represented symbolically by the diagram

$$\underset{2s}{\uparrow\downarrow} \quad \underset{2p}{\uparrow \ \uparrow \ \uparrow} \text{ (hybridize)} \rightarrow \underset{sp}{\uparrow\downarrow \ \uparrow} \qquad \underset{p}{\uparrow \ \uparrow}$$

In our example, one of the hybrid sp orbitals holds a lone pair and the other contributes to the formation of a sigma bond. The two remaining p orbitals, each with one unpaired electron, participate in pi bonding to form two additional bonds in the triple bond. Generally, pi bonds are weaker than sigma bonds and act to make molecular geometries more rigid with respect to rotation about the bonding axis.

Topic Test 2: Hybridization and Bonding

True/False

1. The carbon atom in CH_4 undergoes sp^3 hybridization.

2. O_2 possesses two pi bonds.

3. Two s orbitals and one p orbital mix in sp^2 hybridization.

Multiple Choice

4. Which of the following molecules possess exactly three sigma bonds?
   a. N_2
   b. H_2O
   c. CO_2
   d. NH_3
   e. CH_3I

5. What hybridization does the oxygen atom in H_2O undergo?
   a. sp
   b. sp^2
   c. sp^3
   d. sp^3d
   e. None of the above

Short Answer

6. Construct a diagram that shows the hybridization that the oxygen atom in formaldehyde (H_2CO) undergoes.

7. Give the hybridization on each carbon atom in the following compound:

$$CH_3—CH=CH—C≡N|$$
$$\quad 1 \qquad 2 \qquad 3 \qquad 4$$

Topic Test 2: Answers

1. **True.** The carbon in CH_4 undergoes sp^3 hybridization to achieve four degenerate bonding orbitals in tetrahedral geometry.

2. **False.** Every double bond is composed of one sigma and one pi bond.

3. **False.** sp^2 hybridization involves mixing *one* s orbital with *two* p orbitals.

4. **d.** The three N—H bonds in NH_3 each consist of a sigma bond.

5. **c.** The oxygen in H_2O undergoes sp^3 hybridization. Two of the hybrid sp^3 orbitals house lone pairs and two of the hybrid sp^3 orbitals form sigma bonds with H atoms.

6. The oxygen in H_2CO is surrounded by two lone pairs, one sigma bond, and one pi bond. Accordingly, it undergoes sp^2 hybridization to accommodate two lone pairs and a sigma bond and uses its remaining p orbital for the formation of a double bond. The O atom hybridization can be shown schematically in the following way:

$$\underset{2s}{\uparrow\downarrow} \qquad \underset{2p}{\uparrow\downarrow \; \uparrow \; \uparrow} \text{(hybridize)} \rightarrow \underset{sp^2}{\uparrow\downarrow \; \uparrow\downarrow \; \; \uparrow} \qquad \underset{p}{\uparrow}$$

7. The hybridization is deduced by considering the number of bonding pairs surrounding each carbon atom. Remember to treat multiple bonds as one bonded pair.

$$C(1): sp^3 \qquad C(2): sp^2 \qquad C(3): sp^2 \qquad C(4): sp$$

TOPIC 3: MOLECULAR ORBITAL THEORY

KEY POINTS

✓ *What is a molecular orbital?*

✓ *How do antibonding orbitals differ from bonding orbitals?*

✓ *What is bond order? How does bond order relate to the formation of single, double, and triple multiple bonds?*

Although hybridization provides a useful explanation of bonding with respect to molecular geometry, the **molecular orbital model** better explains the energy states of orbitals involved in bonding. In this theory, atomic orbitals are combined to form **molecular orbitals** that house electrons shared between two atoms. In many ways, molecular orbitals have similar characteristics to atomic orbitals. For example, they hold a maximum of two electrons with opposing spins and have discrete energies. However, these orbitals are associated with a bonded pair of atoms rather than with an individual element. Molecular orbital theory provides an especially useful description of bonding in **homonuclear diatomic molecules**, which are molecules comprised of two identical atoms. Molecular orbitals for these molecules are constructed by combining identical orbitals from each atom (1s ↔ 1s, 2s ↔ 2s, 2p ↔ 2p, etc.). Two molecular orbitals are formed from the interaction of each pair of atomic orbitals because atomic orbitals can either be combined constructively or destructively. These interactions are a consequence of the wave-like properties of electrons. The resulting molecular orbitals have energies and shapes defined by the atomic orbitals from which they are derived. As an example, consider the bonding in H_2. Interaction of the 1s orbital in each H atom gives rise to a **sigma bonding molecular orbital ($\sigma_{1s} = 1s + 1s$)** and a **sigma antibonding molecular orbital ($\sigma_{1s}^* = 1s - 1s$)**, where * denotes an antibonding orbital. The energies of these orbitals are shown schematically in **Figure 7.2**. The sigma bonding molecular orbital concentrates electron density between the two H atoms and is lower in energy than the 1s hydrogen atomic orbital. Therefore, electrons in the sigma bonding molecular orbital act to stabilize the molecule and thus hold the atoms in place. In contrast, the σ^* antibonding molecular orbital is higher in energy than the 1s atomic orbital and decreases electron density between atoms. Therefore, electrons residing in the σ^* antibonding molecular orbital have a destabilizing effect on bonding. Similar to determining electron configurations in atoms, ground state electron configurations for molecules are determined by adding electrons to orbitals in order of lowest to higher energies. Therefore, the two electrons in ground state H_2 reside in the σ_{1s} molecular orbital. This electron configuration can be shown symbolically by the notation **H_2: $(\sigma_{1s})^2$**, where occupied molecular orbitals are written in order of increasing energy and electrons are designated by superscripts. In molecular orbital theory, the stability of a bond is characterized by its **bond order**. Bond order is defined as one half the difference between the number of electrons in bonding orbitals and those in antibonding orbitals:

Bond order = 1/2[(no. of e⁻ in bonding orbitals) – (no. of e⁻ in antibonding orbitals)].

In the case of H_2, the bond order is determined by the following calculation: B.O. = 1/2 (2 − 0) = 1. A positive value of bond order indicates the number of bonds formed (1 = single bond, 2 = double bond, etc.) and negative values or zero indicate unfavorable bonding conditions. Noninteger bond orders are also possible.

Covalent bonding in the period 2 homonuclear diatomic molecules can also be explained accurately using molecular orbital theory. The two 2s orbitals combine in a fashion similar to the 1s

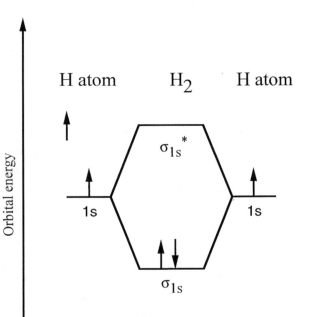

Figure 7.2. Orbital energies of the σ_{1s} and σ_{1s}^* molecular orbitals of H_2.

case and result in σ_{2s} and σ_{2s}^* orbitals. Similarly, mixing the six degenerate p orbitals, three from each atom, results in a total of six molecular orbitals:

$$\sigma_{2p}, \; \sigma_{2p}^*, \; \pi_{2p}, \; \pi_{2p}, \; \pi_{2p}^* \quad \textbf{and} \quad \pi_{2p}^*$$

The energies of these orbitals are shown schematically in **Figure 7.3**. σ_{2p} and σ_{2p}^* orbitals are sigma molecular orbitals and are oriented along the internuclear axis. The remaining four molecular orbitals, π_{2p}, π_{2p}, π_{2p}^*, and π_{2p}^*, are called **pi (π) molecular orbitals** and concentrate electron density above and below the internuclear axis. The relative energies of the 2p molecular orbitals vary with position in the periodic table because of the strong interaction between 2s orbitals with the 2p orbitals in B_2, C_2, and N_2. Figure 7.3 summarizes the energies of the period 2 diatomic molecular orbitals. Given the appropriate molecular orbital energies, the electron configuration of any second-row homonuclear diatomic molecule can be determined by adding electrons into the available molecular orbitals in the order of lowest to highest energies. For example, the electron configuration of O_2 is shown schematically in **Figure 7.4** and can be written in the following manner: $\mathbf{O_2:}$ $(\sigma_{1\sigma})^2 \, (\sigma_{1s}^*)^2 \, (\sigma_{2s})^2 \, (\sigma_{2s}^*)^2 \, (\sigma_{2p})^2 \, (\pi_{2p})^4 \, (\pi_{2p}^*)^2$.

In addition to bond order, the molecular orbital description of bonding accurately predicts the **parity** of homonuclear diatomic molecules. Parity indicates the presence or absence of unpaired electrons in a molecule. Molecules with unpaired electrons in their molecular orbitals are called **paramagnetic** because they are attracted into an inducing magnetic field. In contrast, molecules without unpaired electrons are repelled from an inducing magnetic field and are called **diamagnetic**. As an example, consider the O_2 diatomic molecule that has been experimental shown to be paramagnetic. Although the simple Lewis structure predicts a complete pairing of electrons, the electron configuration in O_2 shown in Figure 7.4 correctly reveals two unpaired electrons in the π_{2p}^* molecular orbitals.

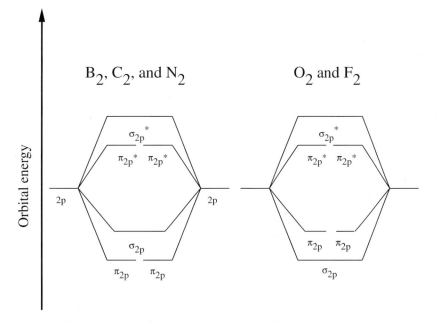

Figure 7.3. Molecular orbital energies for the period two diatomic molecules.

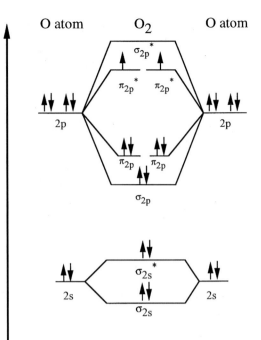

O atom O_2 O atom

Orbital energy

Figure 7.4. Molecular orbital energy-level diagram for the ground state of the O_2 molecule.

Topic Test 3: Molecular Orbital Theory

True/False

1. He_2 has a bond order of 2.

2. Each molecular orbital can accommodate two electrons of opposing spin.

3. The energy of the σ_{2s}^* is always lower than that of the σ_{2s}.

Multiple Choice

4. Which of the following homonuclear diatomic molecules has a bond order equal to three?
   a. H_2
   b. N_2
   c. He_2
   d. F_2
   e. None of the above

5. Which one of the following molecules is expected to be paramagnetic?
   a. H_2
   b. O_2
   c. N_2
   d. F_2
   e. None of the above

Short Answer

6. Arrange the following diatomic molecules in order of increasing bond order: Ne_2, N_2, H_2, and C_2.

Topic Test 3: Answers

1. **False.** The bond order of He_2 is equal to zero (**B.O. = 1/2 [(no. of e⁻ in bonding orbitals) − (no. of e⁻ in antibonding orbitals)] = 1/2 [(2) − (2)] = 0**.

2. **True.** Similar to atomic orbitals, molecular orbitals can accommodate only two electrons. This is a direct consequence of the Pauli exclusion principle (Chapter 5).

3. **False.** The antibonding σ_{2s}^* is higher in energy than the σ_{2s} orbital and results in a decrease in electron density between the bonding atoms.

4. **b.** The bond order of N_2 is calculated using the equation **B.O. = 1/2 [(no. of e⁻ in bonding orbitals) − (no. of e⁻ in antibonding orbitals)] = 1/2 [(10) − (4)] = 3**.

5. **b.** The electron configuration of O_2 (Figure 7.4) indicates two unpaired electrons in the π_{2p}^* molecular orbitals. Thus, it exhibits paramagnetic properties.

6. $Ne_2 < H_2 < C_2 < N_2$. The ordering of elements reflects the electron configurations of each diatomic molecule. Remember to include the influence of s–p interactions on the relative energies of molecular orbitals.

DEMONSTRATION PROBLEM

Using molecular orbital theory, give the ground state electronic configurations of C_2 and F_2. Calculate the bond order expected in each and characterize each molecule as diamagnetic or paramagnetic.

Solution

Using the appropriate relative molecular orbital energies the configurations can be determined for C_2 (12 electrons) and F_2 (18 electrons):

C_2: $(\sigma_{1\sigma})^2 (\sigma_{1s}^*)^2 (\sigma_{2s})^2 (\sigma_{2s}^*)^2 (\pi_{2p})^4$: **diamagnetic**

B.O. = 1/2[(no. of bonding e⁻) − (no. of antibonding e⁻)] = 1/2[(8) − (4)] = 2.

F_2: $(\sigma_{1\sigma})^2 (\sigma_{1s}^*)^2 (\sigma_{2s})^2 (\sigma_{2s}^*)^2 (\sigma_{2p})^2 (\pi_{2p})^4 (\pi_{2p}^*)^4$: **diamagnetic**

B.O. = 1/2[(no. of bonding e⁻) − (no. of antibonding e⁻)] = 1/2[(10) − (8)] = 1.

Chapter Test

True/False

1. Covalent bonds are formed via a mechanism involving cation–anion attraction.

2. Molecular orbital theory explains covalent bonding in terms of hybrid orbitals.

3. The electron geometry of sulfur in SO_2 is linear.

4. Bond formation always liberates energy to the surroundings.

5. F_2 is a homonuclear diatomic molecule that exhibits diamagnetic properties.

Multiple Choice

6. How many lone pairs are in the Lewis dot structure of CO_2?
   a. 0
   b. 1
   c. 2
   d. 3
   e. 4

7. What is the hybridization state of the central P atom in PH_3?
   a. sp
   b. sp^2
   c. sp^3
   d. sp^3d
   e. None of the above

8. Which one of the following homonuclear diatomic molecules is expected to possess a single unpaired electron in its ground state electronic configuration?
   a. B_2
   b. C_2
   c. N_2
   d. H_2
   e. None of the above

9. What is the molecular geometry of CHI_3?
   a. Linear
   b. Tetrahedral
   c. Bent
   d. Octahedral
   e. Trigonal planar

Short Answer

10. What is the bond order of ground state N_2?

11. Choose the polar molecules of the following: H_2O, O_2, CS_2, CF_4, and BrCl.

12. What is the hybridization of the central carbon atom in formaldehyde (H_2CO)?

13. How many sigma bonds, pi bonds, and lone pairs are in the Lewis structure of formaldehyde (H_2CO)?

14. What is the electron configuration of ground state O_2^- atoms (hint: use the molecular orbitals for O_2)?

15. What is the formal charge of oxygen in N_2O?

16. Give the hybridization on each atom in the following compound:

$$|N\equiv C - \overset{\overset{\displaystyle /\overset{\displaystyle O}{}\backslash \; 5}{\|}}{C} - CH_3$$
$$ 1 \quad\;\; 2 \quad\;\; 3 \quad\;\; 4$$

Chapter Test Answers

1. **False**

2. **False**

3. **False**

4. **True**

5. **True**

6. **e**   7. **c**   8. **a**   9. **b**   10. 3

11. H_2O and BrCl

12. sp^2

13. three sigma bonds, one pi bond, and two lone pairs

14. $(\sigma_{1\sigma})^2 \, (\sigma_{1s}^*)^2 \, (\sigma_{2s})^2 \, (\sigma_{2s}^*)^2 \, (\sigma_{2p})^2 \, (\pi_{2p})^4 \, (\pi_{2p}^*)^3$

15. 0

16. N(1): sp, C(2): sp, C(3): sp^2, C(4): sp^3 and O(5): sp^2

Check Your Performance

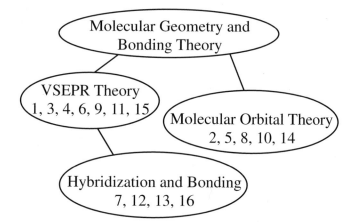

Use this chart to identify weak areas, based on the question numbers you answered incorrectly in the chapter test.

Midterm Exam

True/False

1. The ground state of an F atom has one unpaired electron.

2. KNO_3 is called potassium nitrite.

3. The oxygen atoms in SO_2 undergo sp^2 hybridization.

4. The molecular formula of iron (III) nitrite is Fe_3NO_2.

5. N_2 is a diamagnetic homonuclear diatomic molecule.

Multiple Choice

6. HCl and $Mg(OH)_2$ react via neutralization as described by the following balanced chemical equation: $2HCl + Mg(OH)_2 \rightarrow 2H_2O$. How many grams of HCl are required to react to completion with 37.1 grams of $Mg(OH)_2$?
   a. 46.4
   b. 1.27
   c. 33.1
   d. 0.411
   e. None of the above

7. The density of room air at 25°C is equal to $6.4 \times 10^{-4}\,g/cm^3$. What is the density of air in units of pounds per liter (1 pound = 453.6 grams)?
   a. 7.2×10^3
   b. 9.0×10^{-2}
   c. 1.4×10^{-3}
   d. 7.4×10^2
   e. 1.7×10^{-8}

8. Which electronic configuration below best represents the ground electronic state of the Fe^{2+} ion?
   a. $[Ar]4s^2\,3d^6$
   b. $[Ar]4s^2\,3d^8$
   c. $[Ar]4s^6$
   d. $[Ar]4s^2\,3d^4$
   e. $[Ar]3d^6$

9. A 4.21-mole sample of N_2 is mixed with 11.2 moles of H_2 and reacts to form NH_3: $N_2 + 3H_2 \rightarrow 2NH_3$. Assuming the gas mixture reacts to completion, how many moles of NH_3 are expected to form?
   a. 8.42
   b. 4.21
   c. 11.2
   d. 22.4
   e. 7.47

10. If the specific heat of liquid H_2O is equal to $4.18\,J/(°C\,g)$, how much energy is required to raise the temperature of a 1.20×10^6 liter swimming pool 10.0°C? (Assume the pool is full and water has a density of $1.00\,g/mL$.)

a. $1.12 \times 10^7 \, \text{kJ}$
b. $5.02 \times 10^7 \, \text{kJ}$
c. $2.92 \times 10^6 \, \text{kJ}$
d. $7.31 \times 10^{-7} \, \text{kJ}$
e. $4.44 \times 10^{12} \, \text{kJ}$

Short Answer

11. Using the principles of molecular orbital theory, what is the bond order of ground state B_2?

12. Name the following compounds: N_2O_5, HNO_3, NF_3, $FeSO_4$, $CaCO_3$, and H_2S.

13. Methane (CH_4) is sequentially oxidized to formaldehyde (CH_2O) and carbon dioxide (CO_2) in the Earth's atmosphere. What is the hybridization state of the central carbon atom in each of these compounds?

14. Rank the following atoms in order of increasing ionization energy: S, Mg, Na, Al, and Cl.

15. A 252-nm photon excites an electronic transition in an O_3 molecule. What is the energy associated with the electronic transition?

16. A 476-gram sample of ethylene (C_2H_4) is burned in an excess of oxygen to give CO_2 and H_2O products. Assuming the reaction goes to completion, what mass of H_2O is expected to form?

17. The heats of formation of CH_3OH, CO_2, and H_2O are -201, -393, and $-242 \, \text{kJ/mol}$, respectively. Calculate the standard state enthalpy change accompanying the following reaction: $2CH_3OH + 3O_2 \rightarrow 2CO_2 + 4H_2O$.

18. Write out the molecular formulas of the following compounds: potassium sulfate, dinitrogen monoxide, lithium peroxide, hydrocyanic acid, iron (III) chloride, potassium sulfide, dihydrogen monoxide, and sulfuric acid.

19. Calculate the heat of formation of ammonium nitrate given that NH_3 and HNO_3 have values of ΔH_f° of -45.94 and $-135.1 \, \text{kJ/mol}$, respectively.

$$NH_{3(g)} + HNO_{3(g)} \rightarrow NH_4NO_{3(s)} \qquad \Delta H^\circ = -184.6 \, \text{kJ}$$

20. Under conditions of 1 atmosphere pressure and 298 K, the number density of a sample of air is 2.46×10^{19} molecules per cm^3. How many molecules are present in a cubic micrometer (μm^3) of air?

21. Calculate the change in temperature expected when a 0.33-gram silicon (Si) microchip in your computer absorbs $0.811 \, \text{J}$ of heat at 22°C. [Molar heat capacity of Si $= 19.7 \, \text{J/(mol K)}$.]

22. The combustion of hydrazine (N_2H_4) yields N_2 and H_2O. Write a balanced chemical equation for the combustion of hydrazine.

23. A monolayer (layer of molecules one molecule deep) of oleic acid ($C_{18}H_{34}O_2$: M.M. $= 282.47 \, \text{amu}$) has an area of $20.0 \, \text{cm}^2$ and a thickness of $1.79 \, \text{nm}$. If the density of oleic acid is $0.895 \, \text{g/mL}$, how many molecules are in the monolayer sample?

24. How many atoms of carbon are contained in 0.37 moles of sucrose ($C_{12}H_{22}O_{11}$)?

25. Given the equation below, describe the energy levels available for an He^+ atom:

$$E = -2.178 \times 10^{-18}J(Z^2/n^2)$$

where Z is the nuclear charge (2 for He^+) and n is the principal quantum number. Calculate the energy associated with the $n = 3$ to $n = 1$ transition for He^+.

26. How many grams of NaBr need to be added to 734 mL of water to make a 1.00 M solution?

27. How many protons, neutrons, and electrons are in one $^{56}_{26}Fe^{2+}$ ion?

28. Calculate the number of moles of CO_2 produced from the reaction of a 12-gram sample of O_2 with 44 g of C_2H_6: $2C_2H_6 + 7O_2 \rightarrow 4CO_2 + 6H_2O$.

29. Give the hybridization of the numbered atoms in the following compound:

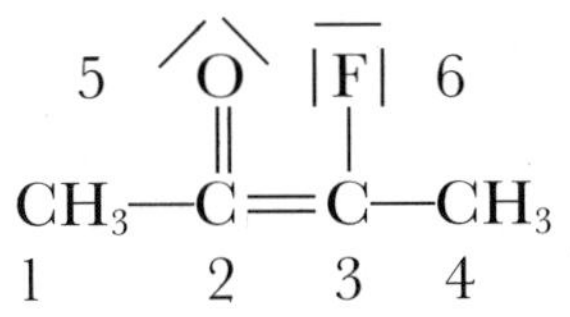

$$
\begin{array}{cccc}
5 & O & |F| & 6 \\
& \| & | & \\
CH_3-C & = & C-CH_3 \\
1 & 2 & 3 & 4
\end{array}
$$

30. Arrange the following elements in order of decreasing atomic radius: F, C, N, Cl, I.

Answers

1. **True**

2. **False**

3. **True**

4. **False**

5. **True**

6. **a**   7. **c**   8. **d**   9. **e**   10. **b**   11. 1

12. Dinitrogen pentaoxide, nitric acid, nitrogen trifluoride, iron (II) sulfate, calcium carbonate, and hydrosulfuric acid.

13. $CH_4(sp^3)$, $CH_2O(sp^2)$, $CO_2(sp)$

14. Na < Mg < Al < S < Cl

15. $7.89 \times 10^{-19}J$

16. 612 grams

17. $-1,352 kJ$

18. K_2SO_4, N_2O, Li_2O_2, HCN, $FeCl_3$, K_2S, H_2O, H_2SO_4

19. $-365.6 kJ/mol$

20. 2.46×10^7 molecules

21. 3.5°C

22. $N_2H_4 + O_2 \rightarrow N_2 + 2H_2O$

23. 6.83×10^{15} molecules

24. 2.7×10^{24} molecules

25. $7.744 \times 10^{-18} J$

26. 75.5 grams

27. 26 protons, 30 neutrons, and 24 electrons

28. 0.21 moles

29. C(1): sp^3, C(2): sp^2, C(3): sp^2, C(4): sp^3, O(5): sp^2, and F(6): sp^3

30. I > Cl > C > N > F

Unit III

Intermolecular Forces and the States of Matter

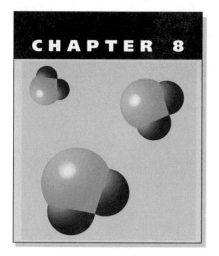

CHAPTER 8

Gaseous State

Compared with other states of matter, relatively few compounds exist as gases under normal conditions. However, the presence of gases at the Earth's surface plays an important role in the Earth's biosphere. We live in a gaseous mixture of primarily N_2 and O_2 gases. Although these gases comprise greater than 99% of the molecules in our atmosphere, trace gas such as H_2O, CO_2, and O_3 also play important roles in establishing a climate at the Earth's surface able to support life. Although the chemical properties of gases vary considerably according to their composition, research over the last 300 years has shown most gases exhibit very similar physical properties. In this chapter we explore what factors govern the physical behavior of gases as observed on the macroscopic level. Ultimately, we develop a theoretical model that explains these physical properties in terms of the motions of the individual particles that comprise gaseous samples.

ESSENTIAL BACKGROUND

- **States of matter (Chapter 1)**
- **SI units and dimensional analysis (Chapter 1)**
- **Kinetic energy (Chapter 4)**

TOPIC 1: GAS PRESSURE AND MEASUREMENT

KEY POINTS

✓ *Pressure is a measure of what property?*

✓ *What are the units of pressure?*

✓ *How is pressure measured?*

Just as a barbell exerts a downward force on the arms of a weight lifter, gas molecules exert a pressure on the container in which they are held. This force arises principally from collisions of translating gas molecules with the walls of their container. **Pressure** is a means of characterizing this force and is equal to the force exerted per unit area: **pressure = force/area**. As one of the most easily measured physical properties, pressure is usually measured using a **barometer**.

The SI unit for force is the newton (N), which is equal to a $kg\,m/s^2$. The SI unit for area is m^2. Inserting both units into the definition above we arrive at the SI unit of pressure: N/m^2 or kg/(m s^2). This derived unit for pressure is called a **pascal (Pa)**. Other pressure units commonly encountered are Torr and atmospheres (atm) and are related to each other by the equality statement: 1 atm = 760 Torr = 1.01325 × 10^5 Pa.

Topic Test 1: Gas Pressure and Measurement

True/False

1. The pressure of a gas is equal to the net force it exerts on its container.

2. There are 760 atm in a 1.00-Torr gas sample.

3. The pressure of an enclosed gas sample is uniform throughout the container.

Multiple Choice

4. What is the pressure of a 352-atm gas sample in units of Torr?
   a. 2.68 × 10^5
   b. 0.463
   c. 352
   d. 2.68 × 10^{-5}
   e. None of the above

5. What is the pressure exerted by a coin (mass equal to 5.00 grams and diameter equal to 1.00 centimeters) resting perfectly flat on a desk? [Hint: force = mass × gravitational constant (9.81 m/sec^2) and area = πr^2.]
   a. 0
   b. 500 Pa
   c. 625 Pa
   d. 1,623 Pa
   e. None of the above

Topic Test 1: Answers

1. **False.** The pressure of a gas sample equals the force per unit area exerted on the walls of a container.

2. **False.** 760 Torr equals 1 atm.

3. **True.** Gases at equilibrium exert a uniform pressure throughout their container.

4. **a.** The conversion from units of atmosphere to Torr is accomplished using the conversion factor 760 Torr/1 atm: 352 atm × (760 Torr/1 atm) = 268 × 10^5 Torr.

5. **c.** The force exerted by the coin can be calculated using the definition of pressure: pressure = force/area. The force in this example is due to gravity: F = (5.00 grams) × (1 kilogram/1,000 grams) × (9.81 m/sec^2) = 4.91 × 10^{-2} kg/(m sec^2). The area of a circle ($A = \pi r^2$) is equal to A = 3.14 × [0.500 × (1 meter/100 centimeters)]^2 = 7.85 × 10^{-5} m^2. The ratio of these values is equal to the pressure: P = 4.91 × 10^{-2} kg/(m sec^2)/7.85 × 10^{-5} m^2 = 625 kg m/sec^2 = 625 Pa.

TOPIC 2: IDEAL GAS LAW

KEY POINTS

✓ *What physical properties does the ideal gas law govern?*

✓ *What is the universal gas constant? What units is it commonly expressed in?*

✓ *How does partial pressure relate to mole fraction?*

Experimentation over the last 300 years has established that the physical properties of gases follow several empirical laws over a wide range of conditions. Exciting discoveries by scientists such as Charles Boyle, Jacques Charles, and Amedeo Avogadro first suggested that the pressure (P), temperature (T), volume (V), and molar amount (n) of a gas sample are correlated to each other. These researchers were able to prove that under most conditions, by holding any two of these variables constant, a simple relationship can be derived for the remaining two. The empirical laws governing the physical behavior of a gas can be combined in the form of the **ideal gas law: $PV = nRT$**, where R is equal to 0.08206 L atm/(K mol) and is called the **universal gas constant**. Given any three variables describing the physical state of a gas sample (P, V, n, or T), the fourth can be determined using the ideal gas law. For example, the pressure of a 4.0-liter gas sample containing 1.0 mole of gas at 298 K can be determined by the following calculation:

$$PV = nRT$$

$$P = nRT/V = [(1.0 \text{ mole}) \times (0.08206 \text{ L atm}/(\text{K mol}) \times (298 \text{ K})]/4.0 \text{ liters} = 6.1 \text{ atm}.$$

Similarly, the volume of 1.000 mole of gas at standard conditions (0°C and 1 atm) can be calculated using the ideal gas law:

$$PV = nRT$$

$$V = nRT/P = [(1 \text{ mole}) \times (0.08206 \text{ L atm}/(\text{mol K})) \times 273.2 \text{ K}]/1 \text{ atm} = 22.42 \text{ liters}$$

This volume (22.42 liters) is called the **molar volume** and is independent of the chemical composition of the gas. The ideal gas law represents a theoretical limit for gas behavior that is approached at low to moderate pressures and high temperatures. Gases that obey the ideal gas law are said to behave ideally. When deviations from ideal behavior are observed, a gas is said to exhibit real gas behavior.

The ideal gas law is also extremely useful for predicting gas behavior upon changing one of its physical properties. For example, if the temperature of a 150-liter gas sample is raised from 298 to 652 K under conditions of constant pressure and molar amount, the final volume of the gas can be calculated by comparing its original and final states. Assuming the gas behaves ideally, the ratio of volume to temperature (V/T) will be a constant: $PV = nRT$ (rearranging) $\rightarrow V/T = nR/P =$ constant. Therefore, initial and final temperatures and volumes can be related to each other by the equation: $V_1/T_1 = V_2/T_2$. The final volume is calculated in the following manner: $V_2 = V_1 \times (T_2/T_1) = 150 \text{ liters} \times (652 \text{ K}/298 \text{ K}) = 328 \text{ liters}$. The ideal gas law can be manipulated to perform similar calculations involving initial and final pressures, temperatures, and molar quantities.

Ideal gases also follow **Dalton's law of partial pressure**, which states that the total pressure of a mixture of gases in a container is equal to the sum of their individual partial pressures: $P_{tot} = P_1 + P_2 + P_3 \ldots$. In this expression, P_x is the partial pressure of gas x. Partial pressure is the pressure each gas would exert if it was alone. The partial pressure of an ideal gas is also related to the fraction of moles of that gas in the total sample. This fraction $(n_x/(n_1 + n_2 + n_3 \ldots))$ is called the **mole fraction** and is equal to the ratio of the partial pressure to the total pressure. Mathematically, this equality can be represented by the following expression: $n_x/n_{total} = P_x/P_{total}$, where x designates the molecule of interest.

Topic Test 2: Ideal Gas Law

True/False

1. The molar volume of any ideal gas is 22.42 liters.

2. All gases behave ideally.

3. The total pressure of a gas mixture containing 45 Torr He and 15 Torr Ar is 60 Torr.

Multiple Choice

4. A gas sample is held at a constant pressure. Initially, the gas occupies a volume of 4.37 liters at 298 K. Assuming the gas behaves ideally, what is the temperature needed to reduce the volume to 3.43 liters?
   a. 380 K
   b. 503 K
   c. 234 K
   d. 304 K
   e. None of the above

5. What is the volume occupied by 32.2 grams of H_2 at standard temperature and pressure (273.2 K and 1.00 atm)?
   a. 361 liters
   b. 22.42 liters
   c. 722 liters
   d. 394 liters
   e. None of the above

Short Answer

6. A cubic centimeter of air contains 2.46×10^{19} molecules of N_2 and O_2. Calculate the number of molecules of O_2 and number of molecules of N_2, assuming that the partial pressures of O_2 and N_2 are 160 and 600 Torr, respectively.

7. At high temperatures, the following decomposition reaction goes to completion: $NH_4Cl_{(s)} \rightarrow NH_{3(g)} + HCl_{(g)}$. If 55.5 grams of NH_4Cl are placed in an evacuated 4.0-liter vessel and heated to 601 K, what is the pressure expected after the decomposition has gone to completion?

Topic Test 2: Answers

1. **True.** The molar volume can be calculated from the ideal gas law for conditions of 1 mole, 273.2 K, and 1 atm.

2. **False.** Ideal behavior is approached at low to moderate pressures and high temperatures.

3. **True.** According to Dalton's law of partial pressures, the total pressure is merely the sum of partial pressures. $P_{total} = 45\,\text{Torr} + 15\,\text{Torr} = 60\,\text{Torr}$.

4. **c.** The required temperature is calculated for conditions of constant pressure and molar amount using the ideal gas law. $V/T = nRT/P = $ constant: $T_2 = V_2 T_1 / V_1 = (3.43\text{ liters} \times 298\,\text{K})/(4.37\text{ liters}) = 234\,\text{K}$.

5. **a.** The sample volume is calculated using the ideal gas law by converting the mass of H_2 to the number of moles. $V = nRT/P = [(32.2\text{ grams}) \times (1\text{ mole}/2\text{ grams}) \times (0.08206\,\text{L atm/mol K}) \times (273.2\,\text{K})]/1\,\text{atm} = 361\text{ liters}$.

6. The partial pressures of O_2 and N_2 are related to their respective mole fractions via Dalton's law of partial pressures.

Calculate mole fractions:

mole fraction $O_2 = 160/760 = 0.211$
mole fraction $N_2 = 600/760 = 0.789$

Calculate number of molecules:

Number of $O_2 = (0.211) \times (2.46 \times 10^{19})$ molecules $= 5.19 \times 10^{18}$ molecules
Number of $N_2 = (0.789) \times (2.46 \times 10^{19})$ molecules $= 1.94 \times 10^{19}$ molecules

7. This problem is solved in two steps involving *reaction stoichiometry* and the *ideal gas law*.

Calculating the moles of NH_3 and HCl generated upon reaction:

$$\text{moles HCl} = (55.5\,\text{grams}) \times (1\text{mole NH}_4\text{Cl}/53.46\,\text{grams})$$
$$\times (1\text{mole HCl}/1\text{mole NH}_4\text{Cl}) = 1.04\,\text{mole HCl}_{(g)}$$

$$\text{moles NH}_3 = (55.5\,\text{grams}) \times (1\text{mole NH}_4\text{Cl}/53.46\,\text{grams})$$
$$\times (1\text{mole NH}_3/1\text{mole NH}_4\text{Cl}) = 1.04\,\text{mole NH}_{3(g)}$$

Calculating the pressure of the vessel:

$$P = nRT/V = [(1.04\,\text{moles} + 1.04\,\text{moles}) \times (0.08206\,\text{L atm/mol K})$$
$$\times (601\text{K})]/4.0\text{ liters} = 26\,\text{atm}$$

TOPIC 3: KINETIC-MOLECULAR THEORY OF GASES

KEY POINTS

✓ *What are the fundamental postulates of the kinetic-molecular theory of gases?*

✓ *How is temperature related to average kinetic energy?*

✓ *What is diffusion? What is effusion?*

✓ *How can the ideal gas law be corrected for real gas behavior?*

Although the ideal gas law describes how gases behave, it provides very little insight as to what phenomena are physically responsible for their behavior. To explain the macroscopic properties of gases, scientists have developed a model of gas behavior called **kinetic-molecular theory**. This model attempts to explain the empirical laws predicting gas behavior in the context of the motions of the individual gas particles comprising the sample. Kinetic-molecular theory is founded on four basic postulates constraining the motions of particles behaving as ideal gases:

1. Gases consist of a large number of particles that are in constant and random motion.

2. The volume of gas particles is negligible when compared with the average spaces between particles.

3. The gas particles exert no forces on each other, and all collisions experienced by gas particles are elastic (i.e., kinetic energy is conserved).

4. The average kinetic energy of a gas sample is proportional to the absolute temperature.

The postulates of kinetic-molecular theory provide a microscopic context in which gas pressure and temperature can be understood. Gas pressure is maintained through the collisions of individual gas particles on the walls of their container. Accordingly, the frequency and strength of molecular collisions determine the magnitude of pressure. The absolute temperature of a gas reflects the average kinetic energy of its gas particles. This relationship can be expressed mathematically by the equation $(KE)_{ave} = 2/3\, RT$. Although a gas sample is characterized by its average kinetic energy, the speeds of individual particles follow a distribution centered about the average value. Therefore, some particles travel at speeds greater than the average speed and other travel at slower speeds. Average kinetic energy is also related to the root mean square (rms) velocity (u_{rms}) of the gas particles. The rms velocity is the speed of a molecule possessing average kinetic energy ($KE = 1/2\, mu_{rms}^2$) and is close in value to the average speed. The two definitions of kinetic energy given above can be combined to yield the following mathematical definition of rms velocity: $u_{rms} = (3\, RT/M)^{1/2}$, where M is the molecular mass, R is the gas constant, and T is temperature.

In addition to describing gas pressure and temperature, kinetic-molecular theory is useful in predicting the rates of molecular effusion and diffusion. Gaseous diffusion is the spontaneous process by which a gas mixes throughout another gas until attaining a uniform distribution. Effusion refers to the process in which a gas leaks out of a vessel through a small hole into a vacuum. Similar to rms velocity, the rates of effusion and diffusion for an ideal gas are inversely proportional to the square root of molecular mass. Therefore, the ratios of effusion and diffusion rates for different molecules are expressed in terms of the appropriate molecular masses:

$$\frac{\text{effusion rate of A}}{\text{effusion rate of B}} = \frac{\sqrt{\text{molecular mass B}}}{\sqrt{\text{molecular mass A}}}$$

$$\frac{\text{diffusion rate of A}}{\text{diffusion rate of B}} = \frac{\sqrt{\text{molecular mass B}}}{\sqrt{\text{molecular mass A}}}$$

Although gas kinetic theory qualitatively explains the physical properties of an ideal gas, its breakdown when describing real world situations is inevitable. Accordingly, scientists have developed a means to correctly predict the behavior of many real gases using a modified form of the ideal gas law. This equation is called the **van der Waals equation** and takes into consideration the space occupied by gas particles and the attractive forces they exert on each other. The van der Waals equation is often expressed in the form:

$$P_{obs} = \frac{nRT}{V - nb} - a\left(\frac{n}{V}\right)^2$$

where a $(atm\,L^2/mol^2)$ and b (L/mol) are experimentally determined factors that vary from gas to gas. nb in this equation corrects for the volume possessed by gas particles themselves and $(n/V)^2$ corrects for attractions between particles. In the limits of low pressure and high temperatures, the van der Waals equation reduces to the ideal gas law.

Topic Test 3: Kinetic-Molecular Theory of Gases

True/False

1. The average kinetic energy of a gas increases with increasing temperature.

2. Real gases behave ideally at extremely low temperatures.

3. He atoms are expected to diffuse more slowly than Ar atoms.

Multiple Choice

4. Which of the following gases is expected to have the highest rate of effusion?
   a. CH_4
   b. N_2
   c. HCl
   d. F_2
   e. CO_2

5. Calculate the pressure of a 3.000-mole sample of N_2 [$a = 1.39$ $(atm\,L^2)/mol^2$ and $b = 0.0391\,L/mol$] occupying a 3.000-liter cylinder at $240\,K$.
   a. $19.10\,atm$
   b. $1.000 \times 10^9\,atm$
   c. $13.00\,atm$
   d. $120.0\,atm$
   e. None of the above

Short Answer

6. Calculate the ratio of the diffusion rates of He and H_2.

Topic Test 3: Answers

1. **True.** Average kinetic energy is linearly proportional to temperature: $KE = 2/3\,RT$. Therefore, an increase in temperature results in a corresponding increase in average kinetic energy.

2. **False.** At low temperatures, gas molecules travel so slowly that intermolecular forces become significant. Therefore, behavior becomes nonideal at very low temperatures.

3. **False.** The diffusion rate is inversely proportional to $\sqrt{M}$. Thus, less massive He atoms will diffuse faster than more massive Ar atoms.

4. **a.** The rate of effusion is inversely proportional to $\sqrt{M}$. Therefore, CH_4 molecules are expected to effuse fastest.

5. **a.** The pressure of N_2 can be calculated using the van der Waals equation

$$P_{obs} = nRT/(V - nb) - a(n/V)^2 = (20.49\,atm - 1.39\,atm) = 19.10\,atm$$

6. H_2 diffuses 1.41 times faster than He. The ratio is calculated in the following manner: rate H_2/rate He $= \sqrt{4.00}/\sqrt{2.02} = 1.41$.

DEMONSTRATION PROBLEM

A 0.149-mole sample of chlorine gas (Cl_2) is placed into a 0.100-liter evacuated cylinder at 310 K. Calculate the pressure of the sample assuming ideal behavior and also using the van der Waals equation ($a = 6.49\,atm\,L^2/mol^2$ and $b = 0.0562\,L/mol$). What is the physical explanation for the difference between the two numbers?

Solution

Use appropriate equations to calculate pressure.

Using the ideal gas law:

$$P = nRT/V = (0.149\ moles)(0.08206\,L atm/(mol K))(310 K)/0.100\ liters = 37.9\,atm$$

Using the van der Waals expression:

$$P = nRT/(V - nb) - a(n/V)^2 = 41.4\,atm - 14.4\,atm = 27.0\,atm$$

The real pressure is lower than that predicted for ideal behavior because molecules exert attractive forces on each other.

Chapter Test
True/False

1. Helium is an ideal gas.

2. A 760-atm gas sample has a pressure of 1,000 Torr.

3. A gas sample containing 79.9 grams of Ar will occupy a volume of 44.8 liters at standard temperature and pressure (STP).

4. Helium diffuses slower than carbon dioxide.

5. The mole fraction of Ne in a gas mixture containing 3.3 atm of Ne and 42.1 atm of N_2 is equal to 0.073.

Multiple Choice

6. The pressure on the top of Mount Everest is measured to be 435 Torr. What is this pressure in atmospheres?

a. 3.31×10^5 atm
b. 1.00 atm
c. 0.572 atm
d. 435 atm
e. None of the above

7. What mass of helium is required to fill a 2,550 liter hot air balloon at standard temperature and pressure?
   a. 455 grams
   b. 213 grams
   c. 910 grams
   d. 4.00 grams
   e. 113 grams

8. How many times faster does methane (CH_4) diffuse than hydrobromic acid (HBr)?
   a. 0.198
   b. 5.05
   c. 0.452
   d. 2.25
   e. None of the above

9. How many moles of H_2O are generated in the combustion of a 3.00-Torr sample of CH_4 occupying a 3.0-liter container at 298 K? Assume an excess of oxygen.
   a. 1.3×10^{-2} grams
   b. 6.0 grams
   c. 9.7×10^{-4} grams
   d. 8.7×10^{-3} grams
   e. 1.7×10^{-2} grams

Short Answer

10. How many liters of O_2 are required to react completely with a 3.0-mole sample of NO at standard temperature and pressure conditions? $2NO_{(g)} + O_{2(g)} \rightarrow 2NO_{2(g)}$

11. At what temperature will a 8.23-mole sample of He occupy a 30.0-liter volume at a pressure of 1.00 atm?

12. Under what temperature and pressure conditions does the ideal gas law break down?

13. A 301.1-Torr gas mixture contains O_2, N_2, and Ar. If the mole fractions of O_2 and N_2 are 0.950 and 0.021, respectively, what is the partial pressure of Ar?

14. A 14.1-gram sample of $CaCO_3$ decomposes to form CaO and CO_2; $CaCO_{3(s)} \rightarrow CaO_{(s)} + CO_{2(g)}$. If the decomposition goes to completion, what volume of gas will be generated at standard pressure and temperature conditions?

15. The pressure at the Earth's surface in sunny Boulder, Colorado is equal to 620 Torr. What is this pressure in units of Pascals?

16. Rank the following gases in order of increasing density at standard temperature and pressure conditions: I_2, Ar, H_2, HBr.

17. How much faster is gaseous He expected to diffuse than gaseous Xe?

Chapter Test Answers

1. **False**

2. **False**

3. **True**

4. **False**

5. **True**

6. **c**   7. **a**   8. **d**   9. **c**

10. 34 liters

11. 44.4 K

12. Low temperatures and high pressures

13. 8.7 Torr

14. 3.16 liters

15. 8.3×10^4 Pa

16. $H_2 < Ar < HBr < I_2$

17. 5.73 times faster

Check Your Performance

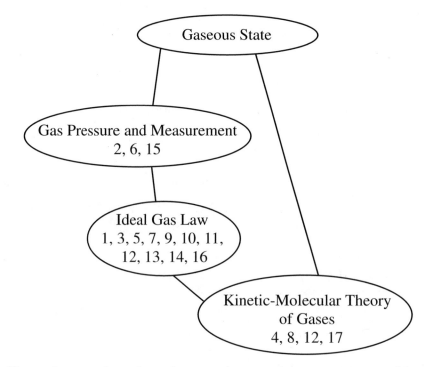

Use this chart to identify weak areas, based on the question numbers you answered incorrectly in the chapter test.

CHAPTER 9

Liquids and Solids

Water is a unique component of our biosphere in that it is present in each of the three states of matter. In its gaseous state, water is an important greenhouse gas that dramatically affects the Earth's climate. Liquid water composes the vast majority of the world's oceans, which supports some of the Earth's most diverse ecosystems. In the form of ice, solid water constitutes the polar ice caps, which are a tremendous reservoir of the Earth's fresh water supply. As shown by the variety of roles water plays on Earth, the chemical and physical properties of a material must relate in some manner to its physical state. In this chapter we consider the physical properties of liquids and solids in light of the fundamental forces that hold these states of matter together.

ESSENTIAL BACKGROUND

- **Covalent and ionic bonding (Chapter 6)**
- **Bond polarity and dipole moments (Chapter 6)**
- **Gas pressure (Chapter 8)**

TOPIC 1: INTERMOLECULAR FORCES

KEY POINTS

✓ *How do intermolecular forces differ from chemical bonds?*

✓ *What are the three main types of intermolecular forces?*

✓ *What is polarizability? How does it vary from molecule to molecule?*

The **condensed states** of matter, solids and liquids, differ from ideal gases in that the individual molecules comprising them interact appreciably to form local aggregates. The interaction between molecules in condensed states can occur by bond formation or by weaker interactions between adjacent molecules. These weaker attractive interactions are called **intermolecular forces** because they exert an attraction between two different molecules rather than within a single molecule. Therefore, intermolecular forces influence the physical state of a material rather than affecting molecular composition. The three most common types of intermolecular forces in order of strongest to weakest are hydrogen bonding, dipole–dipole interactions, and London dispersion forces.

A molecule that possesses a dipole moment exerts electrostatic forces on other polar and nonpolar molecules in its local vicinity. Electrostatic attraction between polar molecules is called a

dipole–dipole interaction and results in an alignment such that the positive and negative ends of adjacent molecules are close to each other. In a condensed phase sample consisting of many molecules, the net alignment of dipoles maximizes the attractive forces of oppositely charged ends (+ . . . −) while minimizing all repulsive interactions (+ . . . + or − . . . −). Dipole–dipole forces are several orders of magnitude weaker than most covalent or ionic bonds and become minimal at large interdipole distances (>10 nm). Molecules containing hydrogen bound to a highly electronegative atom tend to participate in particularly strong dipole–dipole interactions called **hydrogen bonds**, which have important effects on the physical properties of liquids and solids. For example, polar HF molecules readily undergo hydrogen bonding as represented by the following structural drawing: F—H · · · F—H, in which · · · represents a hydrogen bond between two H—F molecules.

Nonpolar species also exert electrostatic forces on other nonpolar species in a condensed-phase sample. Occasionally, the electrons moving about an atom assume a nonsymmetrical electron distribution. The partial charge established in this instantaneous dipole is able to induce similar dipoles in adjacent nonpolar atoms or molecules. Momentary attractive forces between instantaneous dipoles are called **London dispersion forces**. These intermolecular forces are considerably weaker than dipole–dipole and hydrogen bond interactions but are important in determining the physical properties of nonpolar solids and liquids. The strength of London dispersion forces among atoms increases with atomic number because larger atoms have a greater probability of establishing an induced dipole moment. This property is called **polarizability** and is primarily related to the number of electrons possessed by an atom and the distance they reside from the nucleus. Accordingly, the strength of London dispersion forces among nonpolar molecules tends to increase with molecular mass.

Topic Test 1: Intermolecular Forces

True/False

1. Atoms in liquid He are held together by dipole–dipole interactions.

2. H_2O molecules readily form hydrogen bonds with each other.

3. London dispersion forces are stronger than most covalent bonds.

Multiple Choice

4. Which of the molecules below would readily form a hydrogen bond with itself in the condensed phase?
   a. CH_4
   b. H_2
   c. NaBr
   d. CH_3OH
   e. None of the above

5. Which of the following substances is expected to exert the strongest London dispersion forces?
   a. He
   b. Ne
   c. Ar

d. Kr

e. Xe

Topic Test 1: Answers

1. **False.** An He atom possesses no permanent dipole. Therefore, it would not participate in dipole–dipole interactions.

2. **True.** Water readily undergoes hydrogen bonding because it contains two H atoms bound to a highly electronegative oxygen atom.

3. **False.** Typical London dispersion forces are several orders of magnitude weaker than covalent bonds.

4. **d.** CH_3OH would be expected to form hydrogen bonds readily because it contains a highly polar O—H bond (oxygen is highly electronegative).

5. **e.** Xe is the most polarizable element given. It is expected to exhibit the strongest London dispersion forces because it has the largest number of electrons and its electrons are farthest away from the nucleus.

TOPIC 2: PHYSICAL PROPERTIES OF LIQUIDS AND SOLIDS

KEY POINTS

✓ *How do intermolecular forces affect the physical properties of a liquid?*

✓ *What is the difference between crystalline and amorphous solids?*

✓ *What are the different types of crystalline solids?*

Liquids are relatively incompressible fluids that possess a definite volume but no definite shape. The behavior of liquids largely reflects a microscopic condition characterized by relatively strong intermolecular forces and significant molecular motion. Many physical properties of liquids reflect the dynamic nature of the intermolecular forces holding particles in place. As a consequence of the intermolecular forces present in the liquid phase, energy is required to increase the surface area of a liquid. **Surface tension** is a quantitative measure of this amount of energy per unit area. For example, water has a large value of surface tension equal to $7.29 \times 10^{-2} J/m^2$ at 20°C. Thus, $7.29 \times 10^{-2} J$ of energy must be supplied to increase the surface area of a water sample by $1 m^2$. **Viscosity** is a measure of a liquid's resistance to flow and is another property that also depends strongly on the intermolecular forces present in a liquid sample. In general, molecules that have strong intermolecular forces, such as polar molecules, possess large values of viscosity and surface tension.

Solids are ridged materials that possess definite volumes and shapes. Solids that exist in highly ordered arrangements of repeating patterns are classified as **crystalline** and those that exhibit considerable disorder in their structure are called **amorphous**. The repeating patterns in a crystalline solid are called **unit cells** and their three-dimensional arrangement makes up a solid **crystalline lattice**. The physical characteristics of unit cells are described in terms of the lengths and angles defining their edges. Although many different lattice arrangements are found in nature, the simplest consists of the **cubic unit cell**, which possesses sides of equal length and angles of 90 degrees.

The physical properties of most solids depend on the chemical nature of the forces, which hold individual particles together. Therefore, many solids can be conveniently classified with respect to the attractive forces that occur in their crystalline or amorphous forms. **Molecular solids** are held together by intermolecular forces such as dipole–dipole attraction, hydrogen bonding, and London dispersion forces. Examples of this type of solid include dry ice (solid CO_2), solid Ne, and crystalline water ice. Solids consisting of large chains of covalently bound atoms are called **covalent-network solids**. Examples of covalent-network solids are the solid forms of elemental carbon, graphite, and diamond, which consist of carbon atoms covalently bound in different geometries. **Ionic solids** are composed of cation and anion pairs held in place by ionic bonds. Often called salts, ionic solids include NaCl, NaBr, KCl, and KBr, which are common constituents of sea salt. **Metallic solids** comprise the final category of solid materials and are entirely composed of metal atoms held together by forces arising from the interactions of delocalized valence electrons distributed throughout the condensed phase. On a microscopic level, a metallic solid can be viewed as an ordered array of metal cations surrounded by a sea of delocalized valence electrons. These delocalized electrons occupy molecular orbitals very closely spaced in energy called bands, which are formed upon bringing several metal atoms together.

Topic Test 2: Physical Properties of Liquids and Solids

True/False

1. Molecular solids are held together by strong covalent bonds.

2. Increasing the surface area of a liquid requires an input of energy.

3. A solid consisting of highly ordered patterns of unit cells is called an amorphous solid.

4. Liquids with small intermolecular forces tend to have large viscosities.

Multiple Choice

5. Which of the following substances can be viewed as cations bonded together by a sea of delocalized electrons?
   a. $H_2O_{(s)}$
   b. $Fe_{(s)}$
   c. $Ne_{(s)}$
   d. KBr
   e. None of the above

6. Which of the following attractive forces occur in molecular solids?
   a. Dipole–dipole attraction
   b. Hydrogen bonding
   c. London dispersion forces
   d. All of the above
   e. None of the above

Short Answer

7. Classify the type of solid and give the forces present in each of the following materials: $He_{(s)}$, $H_2O_{(s)}$, $Ag_{(s)}$, $CO_{2(s)}$, and $NaNO_{3(s)}$.

Topic Test 2: Answers

1. **False.** Molecular solids are held together by intermolecular forces such as dipole–dipole interactions, hydrogen bonding, and London dispersion forces.

2. **True.** A liquid resists changes that result in an increase in its surface area. Thus, energy must be supplied to increase its surface area.

3. **False.** An amorphous solid is characterized by disorder in the positions of the molecules and atoms that comprise it.

4. **False.** The presence of strong intermolecular forces in a liquid increases a substance's viscosity.

5. **b.** Fe is a metallic solid. Thus, it can be viewed as cations surrounded by a sea of delocalized valence electrons.

6. **d.** Molecular solids are held together by intermolecular forces that include dipole–dipole interactions, hydrogen bonds, and London dispersion forces.

7. $He_{(s)}$: molecular solid, London dispersion forces. $H_2O_{(s)}$: molecular solid, hydrogen bonds. $Ag_{(s)}$: metallic solid, forces arising from the presence of delocalized valence electrons. $CO_{2(s)}$: molecular solid, dipole–dipole attraction. $NaNO_{3(s)}$: ionic solid, ionic bonds.

TOPIC 3: CHANGES OF STATE

KEY POINTS

✓ *What is the heat of vaporization? What is the heat of fusion?*

✓ *What is vapor pressure? How is this property explained on a molecular level?*

✓ *How are changes of state predicted using a phase diagram?*

A **change of state** or **phase transition** is the transformation, which occurs when a substance is converted from one physical state to another. **Melting** or **fusion** is the transformation of a solid to a liquid, for example, $H_2O_{(s)} \rightarrow H_2O_{(l)}$. **Freezing** refers to the reverse process that converts a liquid to a solid, for example, $H_2O_{(l)} \rightarrow H_2O_{(s)}$. **Vaporization** is the phase change of a solid or liquid to a gas, for example, $H_2O_{(l)}$ (or $H_2O_{(s)}$) $\rightarrow H_2O_{(g)}$, and its reverse process, for example, $H_2O_{(g)} \rightarrow H_2O_{(l)}$, is called condensation. The process by which a solid vaporizes directly to form a gas is given the name **sublimation** and its reverse process is called **deposition**.

On a microscopic level, every phase change is characterized by rearrangements in the attractive forces present in a given state and a corresponding change in energy. For example, energy must be supplied to vaporize a liquid water sample. This energy is needed to overcome the intermolecular forces, which hold liquid water molecules together in the condensed phase. Conversely, the condensation of water vapor into the liquid phase releases energy corresponding to the energy of hydrogen bond formation. The flow of energy during any change of state may be quantified in terms of the change in enthalpy. The **heat of vaporization (ΔH_{vap})** is the enthalpy change accompanying the transition between liquid and gaseous states. For example, the vaporization of a mole of liquid water requires the input of 40.66 kJ of heat and can be expressed in the following thermodynamic equation: $H_2O_{(l)} \rightarrow H_2O_{(g)}$ $\Delta H_{vap} = 40.66$ kJ. Similarly, the **heat of fusion (ΔH_{fus})** is the enthalpy change experienced during a phase transition between solid and liquid states.

Changes in phase are also governed by kinetic factors. For example, if a liquid sample of benzene (C_6H_6) is placed in an evacuated closed container at 20.0°C, spontaneous vaporization is observed by a steady increase in reactor pressure. However, after a short time interval, the pressure will achieve a constant value corresponding to the vapor pressure of benzene. At this point the system exists in a state of dynamic equilibrium between vaporization and condensation processes. Under these conditions, the rate of vaporization equals the rate of condensation and the pressure above the liquid reflects the substance's vapor pressure. Because vaporization involves overcoming the intermolecular forces present, the vapor pressure of a liquid increases systematically with increasing temperature or available kinetic energy. The vapor pressures of a liquid at two different temperatures are related to each other by enthalpy of vaporization and are conveniently expressed in the **Clausius Clapeyron equation**:

$$\ln\left(\frac{P_1}{P_2}\right) = \frac{\Delta H_{vap}}{R}\left(\frac{1}{T_2} - \frac{1}{T_1}\right)$$

In this expression, R is the universal gas constant, P is the vapor pressure, and T is temperature. Therefore, given ΔH_{vap} and a liquid's vapor pressure at one temperature, it is possible to predict its vapor pressure at any other temperature.

The temperature at which a liquid's vapor pressure equals the external pressure acting on the liquid surface is called the **boiling point**. Because a substance's vapor pressure strongly depends on its heat of vaporization, boiling point is directly related to the strength of a liquid's intermolecular forces. Liquids exhibiting strong intermolecular attractions possess larger boiling point temperatures than liquids held together by weak intermolecular forces. The temperature at which a liquid changes state to form a solid is called the **freezing point** or **melting point** and may also be related to the heat of fusion associated with a given phase change.

Vapor pressures, boiling points, and melting points for a given substance are often expressed graphically in the form of a **phase diagram**. This diagram shows what equilibrium conditions are established between the three states of matter as functions of ambient temperature and pressure. It is important to note that a phase diagram describes the physical states of pure substances in a closed system at equilibrium. As an example, the phase diagram for H_2O is shown in **Figure 9.1** and is characterized by three curves, which separate the graph area into solid, liquid, and gas regions. In each region, the state indicated is thermodynamically stable. The individual points making up the curves designate ambient conditions in which two phases coexist at equilibrium. Curves separating solid–gas and liquid–gas regions also indicate the equilibrium

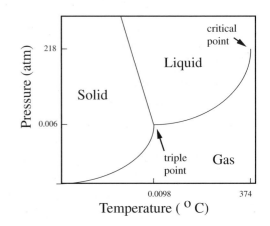

Figure 9.1. Phase diagram for water.

vapor pressures for solids and liquids as a function of temperature. The point at which all three curves intersect is called the **triple point** and represents conditions at which all three states are present at equilibrium. At some temperature the curve separating liquid and gas stability regions comes to an abrupt ending. This temperature is called the **critical temperature**. Above this temperature a vapor cannot be liquefied no matter what pressure is applied and liquid–vapor distinctions cease.

Topic Test 3: Changes of State

True/False

1. The vapor pressure of a liquid decreases with increasing temperature.

2. H_2S is expected to have a lower boiling point than CF_4 at a given pressure.

3. Freezing a sample of liquid water releases heat to the surroundings.

4. Vaporization is the phase change transforming a gas to a liquid.

Multiple Choice

5. Which of the following represents the expected order of boiling point temperatures for the following materials: HF, CH_4, CH_2O, and Ar?
   a. $HF < CH_4 < Ar < CH_2O$
   b. $CH_4 < HF < CH_2O < Ar$
   c. $Ar < CH_2O < CH_4 < HF$
   d. $Ar < CH_4 < CH_2O < HF$
   e. $CH_2O < Ar < CH_4 < HF$

6. Liquid water has a vapor pressure of 20.0 Torr at 298 K and a ΔH_{vap} equal to 40.7 kJ/mol. What is the vapor pressure of water at 273 K?
   a. 4.44 Torr
   b. 30.1 Torr
   c. 5.12×10^{-4} Torr
   d. 21.8 Torr
   e. 18.3 Torr

7. Which of the following phase changes liberate energy to the environment?
   a. Condensation
   b. Sublimation
   c. Vaporization
   d. Melting
   e. All of the above

Short Answer

8. Calculate the heat of vaporization of carbon disulfide (CS_2) given that its vapor pressure is 1.00 atm at 319 K and 0.491 atm at 298 K.

Topic Test 3: Answers

1. **False.** At higher temperatures, the molecules in a sample possess more kinetic energy and more easily overcome the intermolecular forces that hold them in the condensed phase.

2. **False.** H_2S has stronger intermolecular forces than CF_4 in the liquid phase: hydrogen bonds vs. London dispersion forces. Therefore, it is expected to have a higher boiling point.

3. **True.** Energy is always liberated to the surroundings upon a phase change from liquid to solid.

4. **False.** Condensation is the process by which a gas becomes a liquid. Vaporization is when a liquid is converted to a gas.

5. **d.** The order of boiling points reflects the strength of the attractive forces, which occur in the liquid phase. Ar (London dispersion forces) < CH_4 (London dispersion forces) < CH_2O (dipole–dipole) < HF (hydrogen bonding).

6. **a.** The vapor pressure is calculated using the Clausius Clapeyron equation. Use the value of R in units of J/(K mol).

   Defining $T_1 = 273$ K and $T_2 = 298$ K:

   $$\ln(P_1/P_2) = \Delta H_{vap}/R\,(1/T_2 - 1/T_1)$$
   $$= \{[(40.7\,\text{kJ/mol}) \times (1{,}000\,\text{J/1\,kJ})]/(8.3145\,\text{J/kmol}) \times (1/298\,\text{K} - 1/273\,\text{K})\}$$

   $$P_1/P_2 = P_1/20.0\,\text{Torr} = 0.222$$
   $$P_1 = 4.44\,\text{Torr}$$

7. **a.** Condensation results in the formation of intermolecular forces. Therefore, this process liberates energy to the surroundings. The other processes listed use energy to overcome intermolecular forces.

8. 26.8 kJ/mol. The heat of vaporization can be determined using the Clausius Clapeyron equation in the following calculation. Remember to use R in units of J/(K mol).

   $$\ln(P_1/P_2) = \Delta H_{vap}/R\,(1/T_2 - 1/T_1)$$

   Rearranging the equation and solving:

   $$\Delta H_{vap} = (\ln(P_1/P_2) \times R)/(1/T_2 - 1/T_1)$$
   $$= (\ln(0.491\,\text{atm/1\,atm}) \times 8.3145\,\text{J/Kmol})/(1/319\,\text{K} - 1/298\,\text{K})$$
   $$= 26{,}800\,\text{J/mol} = 26.8\,\text{kJ/mol}$$

DEMONSTRATION PROBLEM

Methyl chloride (CH_3Cl) is a compound produced in abundance by phytoplankton in the world's oceans. Calculate the vapor pressure of methyl chloride at the average ocean surface temperature of 288 K, assuming it has a vapor pressure of 101 Torr at 210 K and an enthalpy of vaporization (ΔH_{vap}) equal to 25.0 kJ/mol.

Solution

The Clausius Clapeyron equation relates a liquid's vapor pressure at two temperatures given the enthalpy of vaporization. In this calculation it is more convenient to use R in units of $J/(K \, mol)$.

Calculating the ratio of the two vapor pressures ($T_1 = 298\,K$ and $T_2 = 210\,K$):

$$\ln(P_1/P_2) = \Delta H_{vap}/R \, (1/T_2 - 1/T_1)$$
$$= \{[(25.0\,kJ/mol) \times (1,000\,J/1\,kJ)]/(8.3145\,J/K\,mol)\} \times (1/210\,K - 1/288\,K) = 3.88$$

Calculating the vapor pressure at 298 K:

$$P_1/P_2 = \exp(3.88) = 48.3$$

$$P_1 = P_2 \times 68.7 = 101\,Torr \times 48.3 = 4.88 \times 10^3 \,Torr$$

Chapter Test

True/False

1. In a liquid sample, Kr atoms are held together by dipole–dipole interactions.

2. The vapor pressure of a liquid does not depend on temperature.

3. Water ice is a molecular solid.

4. CF_4 is expected to have a higher boiling point than HF.

5. Liquids with small intermolecular forces tend to have large values of surface tension.

Multiple Choice

6. Which of the following substances is expected to possess the largest vapor pressure at a given temperature?
   a. $H_2O_{(l)}$
   b. $Ne_{(l)}$
   c. $H_2S_{(l)}$
   d. $NH_{3(l)}$
   e. $CO_{2(s)}$

7. Which bonding interaction occurs in a solid sample of KI?
   a. Metallic bonding
   b. Covalent bonding
   c. Ionic bonding
   d. Hydrogen bonding
   e. London dispersion forces

8. The process by which a solid is transformed into a gas is called
   a. sublimation.
   b. condensation.
   c. fusion.
   d. melting.
   e. liberation.

9. How much energy is released to the surroundings when 37.5 grams of gaseous water is condensed into the liquid state? $H_2O_{(l)} \rightarrow H_2O_{(g)}$ $\Delta H_{vap} = 40.66\,kJ/mol$
   a. $40.66\,kJ$
   b. $152\,kJ$
   c. $1.08\,kJ$
   d. $19.5\,kJ$
   e. $84.7\,kJ$

10. Which of the following compounds undergoes hydrogen bonding?
    a. CH_3OH
    b. CF_2O
    c. H_2
    d. CO_2
    e. CH_3Cl

Short Answer

11. Calculate the vapor pressure of water at $230\,K$ given that it possesses a vapor pressure of $20.0\,Torr$ at $298\,K$ and $\Delta H_{vap} = 40.66\,kJ/mol$?

12. Which of the following solids are held together by dipole–dipole interactions or hydrogen bonds? $H_2O_{(s)}$, $CH_{4(s)}$, $Al_{(s)}$, $Ca(OH)_{2(s)}$, $Ne_{(s)}$, and $NH_{3(s)}$.

13. Rank the following materials in order of increasing melting point temperature: $NaCl_{(s)}$, $Ar_{(s)}$, $CH_{4(s)}$, and CH_2O.

14. Which phase of water is expected to be stable at equilibrium under conditions of $200\,K$ and 1.00 atmosphere (Hint: See Figure 9.1)?

15. Which of the following phase changes are endothermic: condensation, vaporization, freezing, melting, and sublimation.

Chapter Test Answers

1. **False**

2. **False**

3. **True**

4. **False**

5. **False**

6. **b**   7. **c**   8. **a**   9. **e**   10. **a**

11. $0.156\,Torr$

12. H_2O and NH_3

13. $Ar_{(s)} < CH_{4(s)} < CH_2O < NaCl$

14. Liquid

15. Vaporization, melting, and sublimation

Check Your Performance

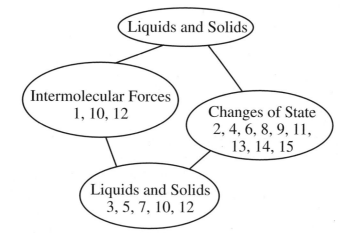

Use this chart to identify weak areas, based on the question numbers you answered incorrectly in the chapter test.

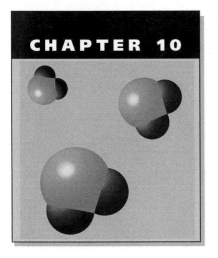

CHAPTER 10

Properties of Solutions

Most materials we encounter on a daily basis are actually mixtures. On a microscopic level, the blood pulsing through our veins, the air we breathe, and the champagne we toast each other with consist of mixtures containing many different atoms or molecules in a variety of physical states. Mixtures that consist of several components uniformly distributed on a molecular level are called **solutions**. Interestingly, the physical properties of solutions often differ considerably from the pure materials of which they are composed. This effect is even observed in incredibly dilute mixtures composed almost entirely of one compound. In this chapter intermolecular forces and energy exchanges associated with solution formation are discussed. Next, the physical properties of solutions are described in terms of their microscopic composition.

ESSENTIAL BACKGROUND

- **Dimensional analysis (Chapter 1)**
- **Enthalpy of reaction (Chapter 4)**
- **Intermolecular forces (Chapter 9)**

TOPIC 1: FUNDAMENTALS OF SOLUTION FORMATION

KEY POINTS

✓ *What is a solution? What processes are involved with solvation?*

✓ *What are the most common ways of expressing concentration?*

✓ *How is energy exchanged between the system and the surroundings during solvation?*

A **solution** is a special type of mixture that consists of two or more components uniformly distributed on the molecular scale. Solutions are known to exist in all three states of matter and often possess one substance in excess called the **solvent**. The other components in a solution, which are present in lesser amounts, are called **solutes**. The properties of solutions are sometimes related to the concentrations of the various solutes in solution. Accordingly, chemists have developed several means of describing the relative abundance of substances in a solution. **Molarity (M)** is the most common unit of concentration and is defined as the moles of solute divided by the volume of solution in units of liters: **M = moles/liters solution**. In addition to molarity, the concentration of a solute can be expressed in terms of mass percent, mole fraction, and molality. Although the definition of each varies, the different ways of relating concentration contain essentially the same information. **Mass percent** is equal to the percentage by mass of the solute in solution: **mass percent = 100 × (mass solute)/(mass solution)**. **Mole fraction (X)** is defined as the ratio of the moles of solute to the total number of moles present in a solution: $X_a = n_a/(n_a + n_b + n_c \ldots)$. **Molality (m)** is the moles of solute per kilogram of solvent: **m = (moles solute)/(kilogram solvent)**.

The process by which a solution is formed is termed **solvation**. We encounter solvation on a daily basis when we dissolve sugar in coffee, add table salt to a pot of soup, or take our daily vitamins. On a molecular level, formation of a solution consists of a rearrangement of the intermolecular forces present in the condensed phase. This rearrangement can be viewed as occurring via three individual processes:

1. Separating the solute sample into individual components;

2. Separating solvent molecules to make room for solute;

3. Forming solute-solvent interactions.

The first two processes involve overcoming solute–solute and solvent–solvent intermolecular forces. Therefore, they consume energy and are endothermic processes: $\Delta H_1 > 0$ and $\Delta H_2 > 0$. The third process involves establishing new intermolecular attractive forces between solute and solvent and is exothermic: $\Delta H_3 < 0$. The overall enthalpy change accompanying solution formation is called the **enthalpy of solution** (ΔH_{soln}) and is equal to the sum of the ΔH values in each step: $\Delta H_{soln} = \Delta H_1 + \Delta H_2 + \Delta H_3$. This summation is expressed graphically in **Figure 10.1**. As shown in Figure 10.1, solution formation can be either endothermic or exothermic depending on the relative magnitudes of ΔH_1, ΔH_2, and ΔH_3.

A large and negative value of enthalpy of solution tends to drive solution formation. However, enthalpy is not the only factor governing dissolution. Solution formation is also driven by the increased disorder that accompanies dissolution. Entropy is a measure of the extent of disorder of a system. When two materials interact to form a uniformly mixed solution, they exist in a

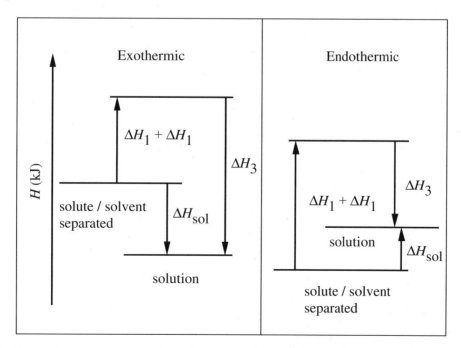

Figure 10.1. Enthapy changes for exothermic and endothermic solution formation.

more disordered state or have a larger entropy. The more disordered state is more probable to exist than the original separated state. Therefore, solution formation is driven by both an exothermic enthalpy of solution and an increase in disorder that accompanies mixing. The thermodynamic considerations needed to predict reaction spontaneity are discussed in terms of the entropy change for a given reaction in Chapter 13. A great simplification of these considerations is found in the observation that *like dissolves like*. This useful rule of thumb indicates that molecules that participate in similar kinds of intermolecular forces tend to dissolve readily in one another and molecules that participate in different types of intermolecular interactions are **immiscible** or do not mix. Thus, polar solutes tend to dissolve in polar solvents and nonpolar solutes tend to dissolve in nonpolar solvents.

Topic Test 1: Fundamentals of Solution Formation

True/False

1. The mole fraction of a solute in a solution is always greater than or equal to 1.

2. A solution formed by dissolving 58.44 grams NaCl in enough water to make exactly 1 liter of solution has a molarity equal to 1.

3. CH_4 is expected to dissolve to a large extent in water.

Multiple Choice

4. A solution containing 21.0 grams of KI per liter of water has a density of $1.014\,g/mL$. What is the molality of the solution?
   a. $1.25 \times 10^{-1}\,mol/kg$
   b. $8.06 \times 10^{3}\,mol/kg$
   c. $1.25 \times 10^{-4}\,mol/kg$
   d. $2.07 \times 10^{-2}\,mol/kg$
   e. $20.7\,mol/kg$

5. Which of the following substances are likely to dissolve readily in water?
   a. $He_{(l)}$
   b. $HCl_{(l)}$
   c. $CF_{4(l)}$
   d. $Xe_{(l)}$
   e. $N_{2(l)}$

6. How many grams of NaOH are needed to make 32.1 liters of a $1.00\,M$ solution?
   a. 1.45×10^{5} grams
   b. 1.28×10^{3} grams
   c. 1.25 grams
   d. 0.800 grams
   e. 3.23×10^{2} grams

Short Answer

7. How many moles of glucose $(C_6H_{12}O_6)$ are present in 4.0 kilograms of a 5.50% mass solution?

Topic Test 1: Answers

1. **False.** Mole fraction is equal to the moles of solute divided by the total number of moles present in a solution: $X_a = n_a/(n_a + n_b + n_c \ldots)$. Therefore, the mole fraction of each component in a solution will always be less than 1.

2. **True.** Molarity is calculated by dividing the moles of solute by the total liters of solution. M = moles NaCl/volume (liters) = [(58.44 grams)/(58.44 g/mol)]/(1 liter) = 1 M.

3. **False.** The rule of thumb for determining whether two materials dissolve to form a solution is *like dissolves like*. In this case, nonpolar CH_4 molecules are unlikely to dissolve appreciably in highly polar H_2O.

4. **a.** Molality is determined in the following calculation using the solution's density: m = moles solute/kilograms solvent = [(21.0 grams)/(166.0 g/mol)]/[(1 liter) × (1.014 g/mL) × (1,000 mL/1 liter) × (1 kilogram/1,000 grams)] = 1.25×10^{-1} mol/kg.

5. **b.** HCl is the only other polar molecule listed. We anticipate HCl will dissolve readily in a water solution because *like dissolves like* or a polar solute dissolves in a polar solvent in this case.

6. **b.** The number of grams required is calculated from the definition of molarity: M = moles solute/liters solution. The molar mass of NaOH equals 39.99 g/mol.

Equations governing concentration:

Molarity = moles NaOH/volume (liter), moles NaOH = grams/molecular mass

Combining equations, rearranging and solving:

Mass NaOH = (M) × (volume (liters)) × (molar mass)

= 1.00 M × 32.1 liters × 39.99 g/mol = 1.28×10^3 grams

7. 1.2 moles $C_6H_{12}O_6$. The moles of solute are determined using the definition of mass percent and the molecular mass of $C_6H_{12}O_6$.

moles of $C_6H_{12}O_6$ = (0.055) × (4.0 kilograms) × (1,000 grams/1 kilogram)
× (1 mole/180.07 grams) = 1.2 moles

TOPIC 2: SOLUBILITY

KEY POINTS

✓ *Why is solubility equilibrium considered a dynamic state?*

✓ *What is a saturated solution? What is a supersaturated solution?*

✓ *Which factors affect a compound's solubility?*

When a solid solute is dissolved in a solvent, its concentration in solution increases steadily until reaching a point at which no more solid will readily dissolve. This amount of solute is referred to as the compound's **solubility** and is usually expressed in units of grams per liter. The solubility of a particular solute is governed by a state of dynamic equilibrium that is achieved between opposing processes of dissolution and crystallization. For example, when $NaBr_{(s)}$ is added to water, the following reversible reaction occurs: $NaBr_{(s)} \rightleftarrows Na^+_{(aq)} + Br^-_{(aq)}$, where the forward reaction represents **dissolution** and the reverse reaction is called **crystallization**. The rate of crystallization depends on the concentration of the solute in solution. Therefore, as more solute dissolves, the rate of crystallization increases. Eventually, the rates of dissolution and crystallization equal each other and no further change in solute concentration is observed. Only appearing static on the macroscopic level, the solution has achieved a state of dynamic equilibrium with the undissolved solute and is called a **saturated solution**. Solute continues to dissolve, but the rate of crystallization is sufficient to prevent any increase in solution-phase concentration. An **unsaturated solution** is one that requires the addition of more solute to achieve saturation. Under some conditions it is possible to dissolve more solute than a compound's solubility would predict. These metastable mixtures are called **supersaturated solutions**. They usually reflect non-equilibrium conditions in which the rate of crystallization is limited by problems associated with forming crystalline solids such as the difficulty involved in achieving the correct lattice orientation.

In addition to solvent composition and structure, solubility also depends on several ambient conditions of the system. The solubility of gas in any solvent increases as its partial pressure over the solvent increases. The relationship between gas pressure and solubility is expressed mathematically by the **Henry's law equation: $P_x = kC$**, where P_x is the gas partial pressure, C is the solution concentration, and k is a constant dependent on the identities of the solvent and solute. Henry's law most accurately predicts the solubilities of gases that do not react or dissociate significantly in solution. Although the solubilities of many solutes have been observed to depend on temperature, predicting the affect of temperature on dissolution is difficult. Most gases exhibit a decreased solubility at higher temperatures. In contrast, many ionic solids show an increased solubility at larger temperatures. However, some ionic compounds, such as $Na_2(SO_4)_2$ and $Ca(OH)_2$, show the reverse behavior. Experimentation is the best way to determine the dependence of a solute's solubility on temperature.

Topic Test 2: Solubility

True/False

1. N_2 is more soluble in water at higher temperatures than at lower temperatures.

2. Supersaturated solutions are in a state of dynamic equilibrium.

3. The solubility of all ionic solids increases with increasing temperature.

4. The rates of crystallization and dissolution are equal in a saturated solution.

Multiple Choice

5. At equilibrium, 1.987 grams of $O_{2(g)}$ are observed to dissolve in 45 liters of water at 20°C and 1 atm. What is the solubility of O_2 in water under these experimental conditions?
   a. 27 g/L
   b. 4.4×10^{-2} g/L
   c. 89 g/L
   d. 23 g/L
   e. None of the above

6. Which of the following molecules is expected to possess the largest solubility in water?
   a. HNO_2
   b. Ar
   c. C_6H_6
   d. N_2
   e. CO_2

Short Answer

7. The solubility of nitrogen gas (N_2) is equal to 1.9×10^{-2} g/L at partial pressure of 1.00 atm. What is the solubility of $N_{2(g)}$ at a reduced pressure of 0.53 atm?

Topic Test 2: Answers

1. **False.** The solubility of a gas tends to decrease as temperature is increased.

2. **False.** A supersaturated solution is in a metastable, nonequilibrium state.

3. **False.** Some ionic solids, such as $Na_2(SO_4)$ or $Ce_2(SO_4)_3$, exhibit a decreasing solubility with increasing temperature.

4. **True.** In a saturated solution, a dynamic equilibrium is established between dissociation and crystallization processes. This results from equal rates of dissociation and crystallization.

5. **b.** The solubility of a solute is the amount that dissolves in a given volume of solute. This is calculated by dividing the mass of O_2 dissolved into solution by the total volume of water: solubility = (1.987 grams O_2)/(45 liters) = 4.4×10^{-2} g/L.

6. **a.** HNO_2 is the most polar of the species listed. Therefore, it is expected to participate in the strongest intermolecular forces when dissolved in a polar solvent such as water. The other compounds listed are nonpolar and do not strongly interact with polar H_2O molecules. Therefore, HNO_2 is expected to possess the largest solubility.

7. 1.0×10^{-2} g/L. The solubility of a gas in a liquid follows Henry's law: $P = kC$. Therefore, the solubility of N_2 can be calculated at any partial pressure knowing the Henry's law constant.

   Calculating the Henry's law constant:

   $$k = P/C = (1.00\,atm)/(1.9 \times 10^{-2}\,g/L) = 53\,atm\,L/g$$

   Calculating the solubility at 0.53 atm pressure:

   $$C = P/k = (0.53\,atm)/(53\,atm\,L/g) = 1.0 \times 10^{-2}\,g/L$$

TOPIC 3: COLLIGATIVE PROPERTIES

KEY POINTS

✓ *What are colligative properties?*

✓ *How does vapor pressure depend on solute concentration?*

✓ *What is osmosis? Why is it a colligative property?*

✓ *How does solute dissociation affect the colligative properties of a solution?*

In many instances, the physical behavior of solutions deviates considerably from the pure substances that comprise them. Experiments have shown that several physical properties of solutions systematically depend on the concentration but not the chemical composition of a solute species present in a solution. These properties are called **colligative properties** and include vapor pressure, boiling point, melting point, and osmotic pressure.

The vapor pressure of a solvent can be dramatically affected by the addition of a nonvolatile or volatile species. **Nonvolatile substances** are those that have a very small vapor pressure and **volatile substances** are those that have significant vapor pressures. When a nonvolatile solute is dissolved in a solvent, the resulting solution always possesses a vapor pressure lower than the original pure solvent. The extent to which a nonvolatile solute depresses vapor pressure is directly proportional to the solute concentration. This relationship is commonly expressed in **Raoult's law: $P_{soln} = X_{solvent}P_{solvent}$**, where P_{soln} is the solution vapor pressure, $X_{solvent}$ is the mole fraction of the solvent, and $P_{solvent}$ is the vapor pressure of the pure solvent. For example, the vapor pressure of a water–glycerin (a nonvolatile solute) solution with a mole fraction of $X_{glycerin} = 0.15$ at 20°C may be predicted using Raoult's law. Given that the vapor pressure of pure water at 20°C is equal to 17.5 Torr, the vapor pressure of the solution is determined via the following calculation: $P_{soln} = X_{H_2O}P_{H_2O} = (1 - X_{glycerin})P_{H_2O}(1 - 0.15) \times (17.5 \text{ Torr}) = 15 \text{ Torr}$.

Raoult's law also predicts the vapor pressures of mixtures of two or more volatile compounds. The vapor pressure of such a mixture is the sum of the expected partial pressures of each volatile component: $P_{soln} = P_A + P_B \ldots = X_A P_A + X_B P_B \ldots$, where subscripts A and B indicate the presence of two volatile compounds. Similar to the concept of an ideal gas, Raoult's law describes the behavior of an **ideal solution**. Solutions tend to behave ideally when they have low solute concentrations and when solute and solvent molecules participate in very similar intermolecular forces. Alternatively, the enthalpy of solution for a given solute and solvent pair can aid in predicting when ideal behavior is observed. Ideal behavior is expected when ΔH_{soln} is close to or equal to 0. The actual solution pressure tends to be larger than Raoult's law predicts when $\Delta H_{soln} < 0$ and lower than Raoult's law predicts when $\Delta H_{soln} > 0$.

Because the addition of a nonvolatile solute systematically decreases the vapor pressure of a solvent, the boiling point and melting point of a substance also exhibits colligative behavior. Raoult's law predicts that the vapor pressure of a solution containing a nonvolatile solute is systematically lower. The boiling temperature of such a solution will be larger than that of the pure solvent. The extent that boiling temperature is elevated is directly proportional to the solute concentration and follows the expression: $\Delta T_b = K_b m$. In this expression, ΔT_b is equal to the difference between solution and solvent boil points ($\Delta T_b = T_b^{solvent} - T_b^{soln}$), K_b is a constant called the **molal boiling point-elevation constant**, and m is the molality of the solution. Similarly, solution melting point is related to solute concentration. However, in this case a decrease in melting temperature is expected upon addition of a nonvolatile solute to a solvent. Melting point depression is observed because melting occurs at the temperature that the vapor pressures of solid and liquid phases are equal. When a nonvolatile solute acts to decrease the solution vapor pressure, this point shifts to lower temperatures. The extent to which melting point is depressed can be predicted using the following expression: $\Delta T_m = K_m m$. In this equation K_m is a constant called the **molal melting-point-depression constant** and m is the concentration of the solute in units of molality.

Osmosis is the process in which a solvent spontaneously flows through a semipermeable membrane. The direction of osmosis always acts to reduce a concentration gradient between both sides of the membrane. The pressure required to prevent the flow of solvent due to osmosis is called **osmotic pressure (Π)**. This property also exhibits colligative behavior. The molarity of the solute (M) is related to the osmotic pressure by the equation $\Pi = \mathbf{MR}T$, where R is the universal gas constant and T is the absolute temperature.

The physical properties of solutions containing dissolved electrolytes exhibit a more complex dependence on solute concentration. For example, consider the boiling point elevation predicted for a 0.10-m solution of HCl in water. The expected melting point depression is predicted by the following calculation: $\Delta T_m = \mathbf{K}_m m = (\mathbf{1.86°C\ kg/mol}) \times (\mathbf{0.20\,m}) = \mathbf{0.37°C}$, because HCl dissociates completely to form one H^+ cation and one Cl^- anion (m = m_{H^+} + m_{Cl^-} = 0.10 m + 0.10 m = 0.20 m). However, the observed melting point depression, $\Delta T_m = 0.35°C$, is slightly less than this prediction. The difference between predicted and observed behavior is related to the electrostatic attractions of the H^+ cations and Cl^- anions in solution. Pairing of oppositely charged ions reduces the net number of independent particles in solution at any given time. The effect of electrostatic interactions among particles is most significant in highly concentrated solutions. To correct for the affect of ion pairing, scientists commonly use an empirically determined correction factor know as the **van't Hoff factor (i)**, which is equal to the number of moles of particles formed in solution per mole of solute dissolved: i = **(moles particles)/(moles solute)**. This correction factor can be conveniently inserted next to the concentration variable in any equation predicting colligative properties to account for ion pairing. For example, the boiling point elevation due to the addition of an electrolyte would be predicted using the following modified equation: $\Delta T_b = i\,K_b m$, where i is the appropriate van't Hoff factor.

Topic Test 3: Colligative Properties

True/False

1. The melting point of a solvent increases upon the addition of an nonvolatile solute.

2. Colligative properties depend on the concentration of solute.

3. Osmotic pressure varies with temperature and solute concentration.

4. Ideal behavior is approached by solutions with large solute concentrations.

Multiple Choice

5. At 20°C the vapor pressures of pure benzene (C_6H_6) and toluene (C_7H_8) are 76 and 21 Torr, respectively. Assuming ideal behavior, calculate the vapor pressure of a solution at 20°C containing 4.7 moles of benzene and 20.1 moles of toluene?
   a. 34 Torr
   b. 2.6 Torr
   c. 97 Torr
   d. 44 Torr
   e. 31 Torr

6. Which of the following aqueous solutions is expected to have the largest vapor pressure?
   a. 0.15 m glucose ($C_6H_{12}O_6$)
   b. 0.15 m HCl

c. 0.15 m NaOH

d. 0.15 m KF

e. They all possess the same vapor pressure.

7. Calculate the melting point expected for a 0.0151 molal aqueous solution of HCl. [$K_m(H_2O) = 1.86°C/m$ and $i = 1.91$]

a. $-0.0281°C$

b. $0.0536°C$

c. $-0.0536°C$

d. $0.0281°C$

e. $-0.0125°C$

Short Answer

8. The vapor pressure of pure H_2O at 25°C is equal to 23.8 Torr. Calculate the vapor pressure of a solution prepared by dissolving 30.0 grams of ethylene glycol ($C_3H_8O_3$: a nonvolatile nonelectrolyte) into 175 grams of water.

Topic Test 3: Answers

1. **False.** The melting point of a solution is always smaller than that of a pure solvent.

2. **True.** Colligative properties depend on concentration and not identity of the solute in solution.

3. **True.** Osmotic pressure is given by the expression $\pi = MRT$, which includes both variables of temperature (T) and concentration (M).

4. **False.** Ideal behavior is approached at *low* solute concentrations.

5. **e.** The vapor pressure of a mixture of two volatile species is calculated using Raoult's law: $P_{soln} = X_A P_A + X_B P_B = (0.19) \times (76 \text{ Torr}) + (0.81) \times (21 \text{ Torr}) = 31 \text{ Torr}$.

6. **a.** $C_6H_{12}O_6$ is the only compound listed that will not undergo dissociation upon addition to water. Therefore, the net concentration of particles in a $C_6H_{12}O_6$ solution will be the smallest, and it is expected to possess the largest vapor pressure.

7. **c.** Melting temperature is a colligative property. Therefore, it is calculated using the melting point depression equation. Recall that the melting point of pure water is equal to 0°C.

Calculating the melting point depression:

$$\Delta T = i K_m m = (1.91) \times (1.86°C/m) \times (0.0151 \text{ m}) = 5.36 \times 10^{-2}°C$$

Calculating the melting point temperature:

$$\Delta T_{soln} = T_m^{solvent} - \Delta T_{soln} = 0°C - 5.36 \times 10^{-2}°C = -5.36 \times 10^{-2}°C$$

8. 23.0 Torr. The vapor pressure of an ideal solution is calculated using Raoult's law: $P_{soln} = X_{solvent} P_{solvent}$. First, the mole fraction of water in the solution must be determined using the definition of mole fraction: $X_{H_2O} = n_{H_2O}/(n_{H_2O} + n_{C_3H_8O_3})$.

Calculating the mole fraction of H_2O:

$$X_{H_2O} = (175 \text{ grams}/18.0\,\text{g/mol})/[(175 \text{ grams}/18.0\,\text{g/mol})$$
$$+ (30.0 \text{ grams}/92.0\,\text{g/mol})] = 0.968$$

Calculating vapor pressure of solution:

$$P_{soln} = X_{solvent}\, P_{solvent} = (0.968) \times (23.8 \text{ Torr}) = 23.0 \text{ Torr}$$

APPLICATION

Pollution can be loosely defined as the unwanted affects of human activity on the Earth's biosphere and climate system. Most pollution consists of emissions of gases, liquids, and solid materials that effect the toxicity of a regional environment. However, some pollution involves unwanted physical changes of materials rather than changing their chemical compositions. Thermal pollution of lakes and rivers is an example of this type of deleterious human impact on the environment. In episodes of thermal pollution, water is drawn off lakes and rivers and used in factories for industrial cooling. Water used for cooling is held in thermal contact with a source of heat but does not become altered chemically. After use as a coolant, the water is returned at a temperature significantly higher than the temperature of its natural source. Because of a decrease in O_2 solubility at higher temperature, the water returning has a significantly lower dissolved oxygen concentration. The thermally polluted water also possesses a lower density than source water and tends to prevent normal mixing that would naturally occur. The ultimate result of thermal pollution is a net decrease in the rate of gaseous oxygen absorption. This in turn has devastating effects on aquatic life. Thermal pollution has the greatest effects on deep lake ecosystems in which a delicate balance is maintained between surface O_2 absorption and the O_2 concentrations at large depths.

DEMONSTRATION PROBLEM

A 0.162-gram sample of an unknown compound is isolated from human blood and is dissolved in 12.0 grams of benzene (C_6H_6). The melting point depression is experimentally determined to be 0.480°C. Assuming that the unknown compound does not dissociate appreciably in solution, what is its molecular mass? (K_f (C_6H_6) = 5.12°C kg/mol)

Solution

The observed melting point depression may be related directly to the molality of the unknown solute via the equation $\Delta T_m = K_f m$. Using the definition of molality [m = (mole solute)/(kg solvent)] and the solute mass, the molecular mass of the unknown solute can be determined.

Calculating the molality of the unknown solute:

$$m = (\Delta T_m)/(K_f) = (0.480°C)/(5.12°C\,\text{kg/mol}) = 9.38 \times 10^{-2}\,\text{mol/kg}$$

Calculating the moles of solute:

moles $= (m) \times$ (mass of solvent (kg))

$$= (9.38 \times 10^{-2} \, \text{mol/kg}) \times (0.012 \, \text{kg}) = 1.13 \times 10^{-3} \, \text{mol}$$

Calculating the molecular mass of unknown:

M.M. $=$ (mass)/(moles) $= (0.162 \, \text{grams})/(1.13 \times 10^{-3} \, \text{mol}) = 143 \, \text{g/mol}$

Chapter Test

True/False

1. The molality and molarity of a solution with a density of $1 \, \text{g/mL}$ are equal.

2. Only solutes with $\Delta H_{\text{soln}} > 0$ dissolve readily into solution.

3. The vapor pressure of a solution is always greater than that of the pure solvent.

4. Molality is equal to the moles solute divided by kilograms of solution.

5. Gases tend to have a reduced solubility at higher temperatures.

Multiple Choice

6. Which of the following aqueous solutions has the lowest melting point?
   a. $0.050 \, \text{m}$ glucose $(C_6H_{12}O_6)$
   b. $0.025 \, \text{NaCl}$
   c. $0.025 \, \text{HNO}_3$
   d. $0.025 \, \text{Ca(OH)}_2$
   e. $0.025 \, \text{HCl}$

7. What is the molarity of a 67.0% H_2SO_4 solution by mass that has a density of $1.61 \, \text{g/mL}$ at $20°C$?
   a. $16.4 \, \text{M}$
   b. $11.0 \, \text{M}$
   c. $5.31 \, \text{M}$
   d. $6.83 \, \text{M}$
   e. None of the above

8. Which of the factors given below affect the solubility of a gas in a liquid solvent?
   a. Temperature
   b. Chemical composition of solvent
   c. Chemical composition of gas
   d. Pressure of gas
   e. All of the above

9. What is the boiling point of a 1.5-m aqueous solution of glucose $(C_6H_{12}O_6$; nonelectrolyte and nonvolatile)? $K_b(H_2O) = 0.512°C/m$
   a. $100.00°C$
   b. $100.77°C$
   c. $100.50°C$
   d. $101.01°C$
   e. $99.23°C$

10. If the solubility of O_2 in rain water is measured to be 9.27×10^{-3} g/L at a partial pressure of 160 Torr, what is the Henry's law constant for O_2 in rain water?
   a. 1.72 atm L/g
   b. 22.7 atm L/g
   c. 0.0579 atm L/g
   d. 15.1 atm L/g
   e. 1.48 atm L/g

Short Answer

11. At 20°C the vapor pressures of pure benzene and toluene are 76 and 21 Torr, respectively. Calculate the mole fraction of benzene in a mixture of benzene and toluene that possesses a vapor pressure equal to 51 Torr at 20°C?

12. How many grams of glycerin ($C_3H_8O_3$; M.M. = 92.03 g/mol) are present in 2.51 liters of a 0.126 M solution?

13. The vapor pressure of water at 25.0°C is equal to 23.8 Torr. How much NaCl needs to be added to 650.0 grams of water to result in a solution with a vapor pressure of 22.9 Torr? Assume a van't Hoff factor for NaCl equal to 2.

14. The solubility of nitrogen gas (N_2) in water is equal to 1.9×10^{-2} g/L at partial pressure of 1.00 atm at 20°C. What pressure of N_2 is required to yield a solubility of 1.5×10^{-1} g/L?

15. To determine the molar mass of an unknown protein, a 1.32×10^{-4} g sample was dissolved in enough water to make a 1.00-mL aqueous solution. If the osmotic pressure of this solution equals 2.29×10^{-3} atm at 25°C, what is the molar mass of the protein? (Assume the protein does not dissociate in solution.)

Chapter Test Answers

   1. **True**

   2. **False**

   3. **False**

   4. **False**

   5. **True**

   6. **d**   7. **b**   8. **e**   9. **b**   10. **b**

   11. 0.55

   12. 29.1 grams

   13. 40.1 grams

   14. 7.9 atm

   15. 1.41×10^3 g/mol

Check Your Performance

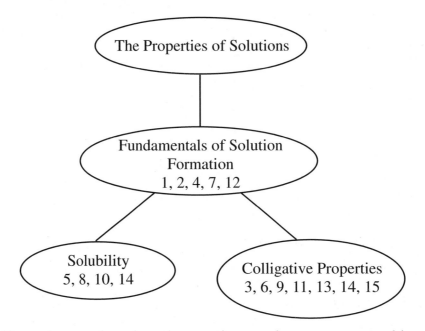

Use this chart to identify weak areas, based on the question numbers you answered incorrectly in the chapter test.

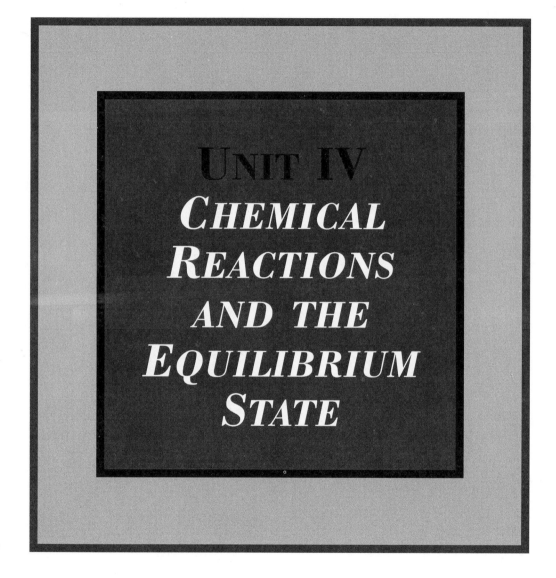

UNIT IV

CHEMICAL REACTIONS AND THE EQUILIBRIUM STATE

Rates and Mechanisms of Chemical Reactions

CHAPTER 11

Chemical reactions proceed over a wide range of times scales, ranging from a few nanoseconds to many centuries. Understanding the times over which chemical reactions occur has important consequences related to all fields of chemistry, including industrial synthesis, biochemistry, forensics, and environmental science. In this chapter we discuss how the rate of a chemical reaction is defined and quantified. We also examine what factors govern reaction rate and develop a microscopic model useful for predicting the rates of many chemical reactions.

ESSENTIAL BACKGROUND

- **Logarithmic and exponential functions**
- **Dimensional analysis (Chapter 1)**
- **Potential and kinetic energy (Chapter 4)**

TOPIC 1: RATE OF REACTION

KEY POINTS

✓ *How is average reaction rate defined?*

✓ *In what units is reaction rate expressed?*

✓ *What is instantaneous reaction rate? How is it determined?*

Chemical kinetics is the study of the rates and mechanisms that characterize a given chemical process. The rate of a chemical reaction is a measure of the speed that reactants are consumed and products are formed. To quantify this characteristic, **reaction rate** is defined as the change in concentration of a reactant or product per unit time and is usually expressed in units of moles per liter per second: mol/(L sec). For example, consider the conversion of ozone (O_3) to oxygen (O_2) in the Earth's atmosphere via the following reaction; $2O_{3(g)} \rightarrow 3O_{2(g)}$. The **average rate** of ozone loss can be determined experimentally by measuring the concentration of O_3 at two different times. Mathematically, the average reaction rate is defined by the equation:

$$\text{rate of } O_3 \text{ loss} = -\frac{[O_3]_2 - [O_3]_1}{t_2 - t_1} = -\frac{\Delta[O_3]}{\Delta t}$$

where $\Delta[O_3]$ and Δt represent the change in ozone concentration and time, respectively. Similarly, the average rate of appearance of the O_2 product over the same time interval is expressed by the expression:

$$\text{rate of } O_2 \text{ appearance} = \frac{[O_2]_2 - [O_2]_1}{t_2 - t_1} = \frac{\Delta[O_2]}{\Delta t}$$

To ensure that reaction rate is always defined as a positive quantity, a negative sign is included when computing loss rates of reacting species but is absent in the product rate expression. In the example above, the chemical equation indicates that three O_2 molecules are formed from the decomposition of every two O_3 molecules. Therefore, the loss rate of O_3 will not equal the production rate of O_2. To equate production and loss rates, the rate of reaction of each species is divided by their respective stoichiometric coefficients:

$$\text{rate of reaction} = -\frac{1}{2}\frac{\Delta[O_3]}{\Delta t} = \frac{1}{3}\frac{\Delta[O_2]}{\Delta t}$$

This equality is useful for relating reactant loss and product formation rates. For example, if $\Delta[O_3]/\Delta t$ is measured to be $-1.0 \times 10^{-5}\,\text{mol}/(\text{L sec})$, the accompanying production rate of O_2 can be determined by rearranging the equation given above and solving:

$$\Delta[O_2]/\Delta t = -3/2\,\Delta[O_3]/\Delta t = -3/2\,(-1.0 \times 10^{-5}\,\text{mol}/(\text{L sec})) = 1.5 \times 10^{-5}\,\text{mol}/(\text{L sec}).$$

The rate of a reaction can also be determined by plotting the concentrations of reactant and product species as a function of time. **Instantaneous reaction rate** is defined as the rate of a reaction at a particular instant in time and equals the slope of a concentration vs. time plot. For example, **Figure 11.1** shows plots of $[O_3]$ and $[O_2]$ vs. time for the reaction discussed above. The instantaneous reaction rate is typically determined by calculating the straight-line tangent of such a plot. Instantaneous reaction rate differs from average rate in that it reflects experimental conditions at one point in time rather than over a finite time interval.

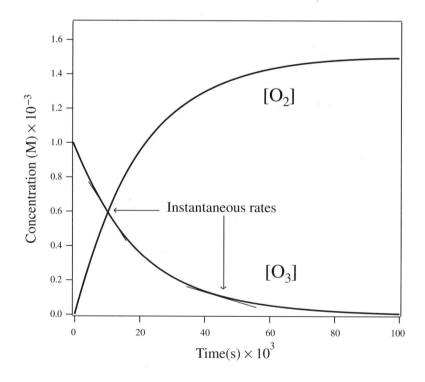

Figure 11.1. Concentration vs. time plot for the reaction $2O_3 \rightarrow 3O_2$.

Topic Test 1: Rate of Reaction

True/False

1. Reaction rate is always expressed in 1/sec units.

2. The rate of loss of a reactant equals the rate of formation of a product.

3. The instantaneous rate is determined by evaluating the straight-line tangent of a plot of reactant concentration vs. time.

Multiple Choice

4. N_2O_5 spontaneously decomposes via the reaction $N_2O_5 \rightarrow NO_2 + NO_3$. If the N_2O_5 concentration is observed to decrease from 0.137 to 0.132 M over a 10.0-minute interval, what is the average loss rate of N_2O_5?
   a. 8.33×10^{-6} mol/(L sec)
   b. -8.33×10^{-6} mol/(L sec)
   c. 5.00×10^{-4} mol/(L sec)
   d. -5.00×10^{-4} mol/(L sec)
   e. 5.00×10^{-3} mol/(L sec)

5. NO_2 reacts with itself to form N_2O_4 according to the following reaction: $2NO_2 \rightarrow N_2O_4$. If the production rate of N_2O_4 is determined to be 2.84×10^{-7} mol/(L sec), what is the accompanying loss rate of NO_2?
   a. 2.84×10^{-7} mol/(L sec)
   b. 1.42×10^{-7} mol/(L sec)
   c. -1.42×10^{-7} mol/(L sec)
   d. 0.71×10^{-7} mol/(L sec)
   e. 5.68×10^{-7} mol/(L sec)

Short Answer

6. Describe the difference between average loss rate and instantaneous loss rate. How are each measured experimentally?

Topic Test 1: Answers

1. **False.** In the context of a chemical reaction, rate is defined as the change in reactant *concentration* per unit time. Because chemists frequently work in molarity units, rate is often expressed in moles per liter per second units [mol/(L sec)].

2. **False.** Reactant loss and product formation rates are related to each other by the stoichiometric coefficients in the balanced chemical reaction.

3. **True.** The easiest way of determining the instantaneous reaction rate is to compute the straight-line tangent at one point on a plot of reactant or product concentration vs. time.

4. **a.** The average rate of N_2O_5 decomposition is determined by dividing the observed change of $[N_2O_5]$ by the time interval of 10 minutes: rate $= -\Delta[N_2O_5]/\Delta t = -([N_2O_5]_{t_2} - [N_2O_5]_{t_2})/\Delta t = -(0.132\,M - 0.137\,M)/[(10\text{ minutes}) \times (60\,\text{sec/min})] = \mathbf{8.33 \times 10^{-6}\,mol/(L\ sec)}$.

5. **e.** The rates of N_2O_4 formation and NO_2 loss are related by the following expression that includes the stoichiometric factors in the chemical equation

$$1/2\,(\Delta[NO_2]/\Delta t) = 1/1(\Delta[N_2O_5]/\Delta t)$$

Solving for $\Delta[NO_2]/\Delta t$:

$$\Delta[NO_2]/\Delta t = 2/1\Delta[N_2O_5]/\Delta t = -2/1(2.84 \times 10^{-7}\,mol/(L\ sec))$$
$$= \mathbf{5.68 \times 10^{-7}\,mol/(L\ sec)}$$

6. Average reaction rate and instantaneous rate differ in the time interval over which they represent the kinetics of a given reaction. Average rate is associated with a finite time interval, whereas instantaneous rate reflects the reaction at one point in time. Average reaction rate is determined by measuring the change in reactant concentration over a measurable time interval. Instantaneous reaction rate is determined by calculating the straight-line intercept at a particular point on a plot of concentration vs. time.

TOPIC 2: DIFFERENTIAL AND INTEGRATED RATE LAWS

KEY POINTS

✓ *What do differential and integrated rate laws describe?*

✓ *What is reaction order? How can it be experimentally determined?*

✓ *What is half-life? How is it calculated?*

As shown in Figure 11.1, the rates of some reactions show dramatic decreases at long reaction times. This behavior is exhibited by a great number of chemical reactions and reflects the dependence of reaction rate on reactant concentration. The **differential rate law** of a chemical reaction describes the relationship between reaction rate and the concentrations of all reactants raised to various powers. For example, the rate of the reaction $NO_2 + NO_3 \rightarrow N_2O_5$ is related to NO_2 and NO_3 concentrations via the following rate law:

$$\text{Rate} = k_{rxn}\,[NO_2]^x\,[NO_3]^y$$

where k_{rxn}, x, and y are each constants that must be determined experimentally. k_{rxn} is commonly referred to as the **rate constant** and x and y are called the **reaction orders**. Experiments have established that the rate law in our example is equal to rate $= k_{rxn}\,[NO_2]\,[NO_3]$, where k has units of $L/(mol\ sec)$ and the reaction orders of NO_2 and NO_3 are each equal to 1. Given the differential rate law and rate constant, the reaction rate for any process can be predicted for every possible condition of reactant concentration.

The **method of initial rates** is one of the simplest ways of determining reaction order. This technique involves measuring the instantaneous reaction rate immediately after a reaction begins for a series of reactant concentration conditions. By comparing the magnitudes of initial rates under differing experimental conditions, the reaction order for each reactant is deduced. For example, consider the following initial rate data collected for the reaction $NO_2 + NO_3 \rightarrow N_2O_5$.

Experiment	Initial [NO$_3$] (mol/L)	Initial [NO$_2$] (mol/L)	Initial Rate (mol/(L sec))
1	0.010	0.005	1.05×10^{-8}
2	0.010	0.010	2.10×10^{-8}
3	0.020	0.010	4.20×10^{-8}

The ratio of the initial rates of experiments 2 and 1 is used to determine the reaction order with respect to NO$_2$ by substituting in initial conditions and solving:

$$\text{Rate 2/rate 1} = \left(k_{rxn}[NO_2]^x[NO_3]^y\right)\Big/\left(k_{rxn}[NO_2]^x[NO_3]^y\right)$$

$$2.10 \times 10^{-8}\,\text{mol(L sec)}/1.05 \times 10^{-8}\,\text{mol(L sec)}$$

$$= \left(k_{rxn}(0.010\,M)^x(0.010\,M)^y\right)\Big/\left(k_{rxn}[0.005\,M]^x[0.010\,M]^y\right)$$

$$2.00 = (0.010\,M)^x\Big/(0.005\,M)^x = (2)^x; \qquad \text{therefore } x = 1$$

Similarly, the ratio of the initial rates of experiments 3 and 2 may be used to determine the order with respect to NO$_3$ in the following manner:

$$\text{Rate 3/rate 2} = \left(k_{rxn}[NO_2]^x[NO_3]^y\right)\Big/\left(k_{rxn}[NO_2]^x[NO_3]^y\right)$$

$$4.20 \times 10^{-8}\,\text{mol(L sec)}/2.10 \times 10^{-8}\,\text{mol(L sec)} = \left(k_{rxn}(0.01)^x(0.020)^y\right)\Big/\left(k_{rxn}[0.010]^x[0.010]^y\right)$$

$$2 = (0.020)^y\Big/(0.010)^y = (2)^y; \qquad \text{therefore, } y = 1$$

Thus, the differential rate law for this reaction is rate = k [NO$_2$][NO$_3$]. Once the differential rate law is known, the value of the rate constant can be determined using the rate and reactant concentration data of any experiment. Using the initial rate data in experiment 1, k is evaluated in the following calculation:

Differential rate law:

$$\text{Rate} = k[NO_2][NO_3]$$

Rearranging the rate law:

$$k = (\text{rate experiment 1})/([NO_2][NO_3])$$

Substituting in reactant concentrations from experiment 1:

$$k = (1.05 \times 10^{-8}\,\text{mol}/(L\,\text{sec}))/[(0.010\,M) \times (0.005\,M)] = 2.10 \times 10^{-4}\,L/(\text{mol s})$$

Although the differential rate law for a given reaction provides useful information pertaining to the dependence of reaction rate on time, often it is more valuable to follow the course of a reaction by monitoring the concentration of a reactant or product. The **integrated rate law** expresses the reactant concentration as a function of time and is derived from the differential rate law. The algebraic forms of the integrated rate law depend on reaction order and are summarized in **Table 11.1** for reactions with orders of 0, 1, and 2. Because reactions of different order possess different integrated rate laws, reaction order can also be deduced by plotting concentration vs. time data. First-order processes yield a straight-line plot of ln[reactant] vs. time. A plot of [reactant] vs. time will show a straight line for zero-order reactions and a plot of 1/[reactant] vs. time will produce a straight line for second-order reactions involving one reactant species. The criteria for determining reaction order graphically are also summarized in Table 11.1.

Table 11.1 Summary of Differential and Integrated Rate Laws for 0, 1, and 2 Reaction Orders

	ZERO ORDER	FIRST ORDER	SECOND ORDER
Differential rate law	Rate = k	Rate = k[X]	Rate = k[X]^2
Integrated rate law	$[X]_t = -kt + [X]_0$	$\ln[X]_t = -kt + \ln[X]_0$	$1/[X]_t = kt + 1/[X]_0$
Linear plot	[X] vs. time	ln[X] vs. time	1/[X] vs. time
Half-life	$[X]_0/2k$	0.693/k	1/k[X]

Half-life is defined as the time required to reduce a reactant concentration to one half of its initial value. Although the reactant concentration follows a different functional dependence with respect to time for each reaction order, the half-life of any process can be related to the rate coefficient and/or initial concentrations of reacting species. Table 11.1 presents the equations that define the half-life for a given reactant depending on the integrated rate law.

The rate laws described in Table 11.1 only deal with simple reactions involving only one type of reactant species. In reality, a great number of chemical reactions involve two or more reacting materials. Although the kinetic behavior of reactions involving several reactants is complex, under certain conditions these reactions exhibit first-order behavior. These experimental conditions are called **pseudo-first-order** conditions and occur when one or more reactants are present in a concentration that greatly exceeds the others. Under pseudo-first-order conditions, the concentrations of reactants in great excess do not change appreciably during reaction. Therefore, the concentrations of such reactants may be regarded as constants in the differential rate law. Therefore, these concentrations do not contribute to defining the overall reaction order. For example, if the concentration of NO_3 greatly exceeds the concentration of NO_2, the differential rate law for the reaction $NO_2 + NO_3 \rightarrow N_2O_5$ can be written as rate = k'_{rxn} [NO_2], where k'_{rxn} is equal to the product of the rate constant and the concentration of NO_3; $k'_{rxn} = k_{rxn}$ [NO_3]. As expected for any first-order reaction, a plot of [NO_2] vs. time would produce a straight line under these conditions.

Topic Test 2: Differential and Integrated Rate Laws

True/False

1. The differential rate law relates reactant concentration with time.

2. The half-life of a first-order process is independent of concentration.

3. A plot of 1/[reactant] vs. time yields a straight line for a zero-order reaction.

Multiple Choice

4. The decomposition of chlorine nitrate ($ClONO_2$) is a first-order process that occurs via the following reaction: $ClONO_2 \rightarrow ClO + NO_2$. If the rate coefficient for this process is equal to $1.53 \times 10^2 \, sec^{-1}$, what is the half-life of $ClONO_2$?
   a. 1.53×10^2 seconds
   b. 2.21×10^2 seconds
   c. 4.53×10^{-3} seconds

d. 6.53×10^{-3} seconds

e. None of the above

5. The following reaction $2NOCl \rightarrow 2NO + Cl_2$ is shown to exhibit zero-order kinetics with a rate coefficient equal to $2.50 \times 10^{-7}\,mol/(L\,sec)$. If the initial concentration NOCl is equal to $0.165\,M$, what is its concentration after a time interval of 22.0 hours?

a. $0.022\,M$

b. $0.159\,M$

c. $0.145\,M$

d. 0

e. $0.112\,M$

Short Answer

6. The following initial rate data were collected for the reaction $2NO + O_2 \rightarrow 2NO_2$.

Experiment	Initial [NO] (mol/L)	Initial [O₂] (mol/L)	Initial Rate (mol/(L sec))
1	0.0050	0.0050	3.05×10^{-8}
2	0.010	0.0050	6.10×10^{-8}
3	0.010	0.010	2.44×10^{-7}

Using the method of initial rates, determine the overall reaction order and rate coefficient for this reaction.

Topic Test 2: Answers

1. **False.** The differential rate law describes the relationship between reaction rate and time.

2. **True.** The half-life of a first-order process is given by the equation $t_{1/2} = 0.693/k_{rxn}$, which does not depend on concentration.

3. **False.** A plot of [reactant] vs. time yields a straight line for zero-order processes.

4. **c.** The half-life of a first-order process is given by the equation $t_{1/2} = 0.693/k_{rxn}$. By inserting the value of the rate coefficient, the half-life is determined in the following calculation: $t_{1/2} = 0.693/k_{rxn} = 0.693/1.53 \times 10^2\,sec^{-1} = \mathbf{4.53 \times 10^{-3}}$ **seconds**.

5. **c.** The integrated rate law for zero-order reactions may be used to calculate the concentration of N_2O_5 after 22 hours. $[N_2O_5]_t = [N_2O_5]_0 - k_{rxn}\,t = 0.165\,M$ $- (2.50 \times 10^{-7}\,mol/(L\,sec)) \times (22\,hours) \times (60\,min/1\,hr) \times (60\,s/1\,min) = \mathbf{0.145\,M}$.

6. **Overall order = 3 and $k_{rxn} = 2.4 \times 10^{-1}\,L/(mol\,sec)$.** The differential rate equation must be determined by using the method of initial rates. Then, the rate constant is determined by substituting the reactant concentrations into the differential rate equation.

   Determining the reaction order with respect to NO:

   $$[\text{Rate 2/rate 1}] = (6.10 \times 10^{-8}\,mol/(L\,sec))/(3.05 \times 10^{-8}\,mol/(L\,sec)) = 2$$
   $$2 = \left[k_{rxn}(0.010\,M)^x\,(0.005\,M)^y\right] / \left(k_{rxn}[0.005\,M]^x\,[0.005]^y\right) = (2)^x,\ x = 1$$

Determining the reaction order with respect to O_2:

$$[\text{Rate 3/rate 2}] = (2.44 \times 10^{-8}\,\text{mol}/(\text{L sec}))/(6.10 \times 10^{-8}\,\text{mol}/(\text{L sec})) = 4$$

$$2 = \left[k_{rxn}(0.010\,\text{M})^x(0.010\,\text{M})^y\right]/\left(k_{rxn}[0.010\,\text{M}]^x[0.005]^y\right) = (2)^x, \; x = 2$$

Determining the reaction rate coefficient:

$$\text{Rate} = k_{rxn}[\text{NO}][\text{O}_2]^2$$

Rearranging equation and solving for condition in experiment 1:

$$k = (\text{initial rate})/\left([\text{NO}][\text{O}_2]^2\right)$$

$$= (3.05 \times 10^{-8}\,\text{mol}/(\text{L sec}))\,0\,/\left[(0.0050\,\text{M}) \times (0.0050\,\text{M})^2\right]$$

$$= 2.4 \times 10^{-1}\,\text{L}/(\text{mol sec}).$$

TOPIC 3: REACTION MECHANISMS AND DYNAMICS

KEY POINTS

✓ *What is an elementary reaction?*

✓ *What determines the overall rate of a complex reaction mechanism?*

✓ *How does reaction rate depend on temperature?*

✓ *How does the presence of a catalyst affect a chemical reaction?*

A balanced chemical equation relates the species consumed during a chemical reaction to the products generated. However, the actual processes that occur during reaction are often more complex than the changes indicated in the chemical equation. Many chemical reactions occur by a series of individual chemical processes called a **reaction mechanism**. For example, the decomposition of ozone in the upper stratosphere actually proceeds through the following mechanism involving two steps:

Step 1:	O_3	$\rightarrow O + O_2$
Step 2:	$O + O_3 \rightarrow O_2 + O_2$	
Net reaction	$2O_3$	$\rightarrow 3O_2$

In this mechanism, the O atom is formed and consumed during the reaction sequence and is called a **reaction intermediate**. The individual steps represented in a reaction mechanism consist of **elementary reactions** that occur in a single event. The number of reactant molecules participating in an elementary reaction determines its **molecularity**. Elementary reactions involving one reactant are called **unimolecular**. Reactions characterized by the collision of two and three reactants are called **bimolecular** and **termolecular**, respectively. Because an elementary reaction reflects a single-step decomposition or collision-induced reaction, the form of its rate law follows the reaction molecularity: **unimolecular = first order, bimolecular = second order,** and **termolecular = third order**. Thus, for elementary reactions the orders of individual reactant species can be directly inferred from their stoichiometric coefficients.

A proposed reaction mechanism must also explain the experimentally observed rate law for a given process. Often, multistep reaction mechanisms possess one elementary step that is considerably slower than the others. In this case, the overall rate of the reaction sequence is limited by the rate of the slowest step. This elementary reaction is designated as the **rate-determining step** because its rate law reflects the rate law of the entire reaction mechanism. For example, in the decomposition of ozone, the rate-limiting step is the first elementary reaction in the mechanism (see above). Thus, the rate of the entire mechanism is expected to follow its differential rate law expression: rate $= k_{rxn}[O_3]$. Occasionally, several slow elementary processes limit the rate of a reaction sequence. In this case, the rate law reflects a combination of the individual rate laws for each limiting step and is usually a complex function of reactant and intermediate concentrations.

In addition to concentration, the rates of most elementary reactions increase as temperature is raised. A microscopic model of chemical reactions, called the **collision theory**, reasonably explains this dependence of reaction rate. The collision theory model postulates that molecules react via collision and qualitatively explains the dependence of reaction rate on both reactant concentration and temperature. Increasing reactant concentrations increases the net number of collisions per unit time between reacting species and ultimately leads to faster rates. Increasing temperature results in a greater average velocity of a reacting sample and thus increases the collision rate. Therefore, a reacting sample can be viewed as a collection of particles continuously colliding with each other to cause reaction. Experimental observations show that only a small fraction of collisions actually have the required energy to actually induce chemical change. This minimum amount of energy is called the **activation energy** and varies from reaction to reaction. The activation energy is often viewed as an energy barrier that must be overcome to allow for the necessary bond rearrangement involved with product formation to occur. The arrangement of atoms at the top of the energy barrier is called the **transition state** and it reflects the geometry of reacting species that will lead to reaction. The relationship between activation energy and the transition state is summarized in the reaction coordinate diagram shown in **Figure 11.2**.

The temperature dependence of a chemical reaction can often be quantitatively predicted using the **Arrhenius equation**: $k_{rxn} = A\, e^{\{-E_a/RT\}}$, which relates the value of the rate constant to temperature (T) and the universal gas constant $(R = 8.314\,J/mol\ K)$. In this expression, A is a variable called the **frequency factor** and is related to the frequency of collisions that have an orientation favorable for reaction. E_a represents the **activation energy** and is usually given in units of kJ/mol. Although A exhibits a slight dependence on temperature, it is usually approximated as a constant. The Arrhenius equation can be represented graphically by plotting ln k vs. $1/T$ to yield a linear plot.

The rates of complex reaction sequences can also be affected by the presence of nonintermediate species called catalysts. The addition of a catalyst always results in an increase in the reaction rate. In addition, catalysts are not consumed or produced during reaction. For example, the rate of ozone decomposition is dramatically increased in the presence of Cl atoms due to the following sequence of reactions:

Step 1:	$O_3 + Cl \rightarrow ClO + O_2$
Step 2:	$ClO + O_3 \rightarrow O_2 + O_2 + Cl$
Net reaction	$2O_3 \rightarrow 3O_2$

In this mechanism, reaction with Cl atom in Step 1 provides a reaction pathway with a lower activation energy than that of the first elementary reaction in the uncatalyzed mechanism. Ultimately, this results in a net increase in the rate of the overall conversion of O_3 to O_2. Catalysts are generally classified as heterogeneous or homogeneous depending on their physical state. A **homogeneous catalyst** is present in the same phase as the reactant molecules, while a **heterogeneous catalyst** exists in a different physical state.

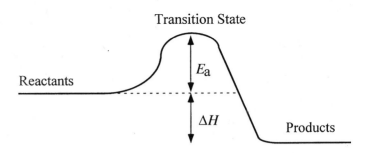

Figure 11.2. Reaction coordinate diagram.

Topic Test 3: Reaction Mechanisms and Dynamics

True/False

1. All unimolecular reactions have first-order rate laws.

2. The fastest chemical reaction in a mechanism is usually the rate-determining step.

3. The rate of an elementary reaction is linearly dependent on temperature.

4. The addition of a catalyst always increases the rate of reaction.

Multiple Choice

5. In the following reaction mechanism, which species acts as a catalyst?
$2NO + O_2 \rightarrow 2NO_2$
$2NO_2 + 2SO_2 \rightarrow 2SO_3 + 2NO$
a. NO
b. NO_2
c. O_2
d. SO_2
e. No catalyst is present.

6. Which of the rate laws given below best represents the reaction $2NO + Br_2 \rightarrow 2NOBr$ that proceeds via the following mechanism?
$NO + Br_2 \rightarrow NOBr_2$ (fast reaction)
$NOBr_2 + NO \rightarrow 2NOBr$ (slow reaction)
a. rate $= k_{rxn} [NOBr_2]$
b. rate $= k_{rxn} [Br_2] [NO]$
c. rate $= k_{rxn} [NOBr_2] [NO]$
d. rate $= k_{rxn} [NO]^2 [Br]$
e. rate $= k_{rxn} [NOBr_2] [NO]^2 [Br_2]$

Short Answer

7. The reaction $OH + CH_3SCH_3 \rightarrow H_2O + CH_3SCH_2$ is characterized by an A factor equal to $9.00 \times 10^{-12} cm^3/(molecule \ sec)$ and an activation energy of $1.94 \times 10^3 J/mol$. Calculate the rate constant for this reaction at $301\,K$.

Topic Test 3: Answers

1. **True.** The order of an elementary reaction reflects the molecularity.

2. **False.** The rate-determining step comprises the slowest elementary reaction in the mechanism.

3. **False.** The temperature dependence of the rate coefficient is described by the Arrhenius equation: $k = A\exp\{-E_a/RT\}$. This is not a linear function.

4. **True.** Addition of a catalyst always increases the reaction rate because it provides a pathway with a lower activation energy.

5. **a.** NO acts as a catalyst because it is used up in Step 1 and reformed in Step 2.

6. **c.** The rate of the overall reaction reflects the rate law for the rate-determining step (Step 2). Because this is an elementary reaction, its rate law is first order with respect to each reactant.

7. $4.14 \times 10^{-12} \text{cm}^3/(\text{molecule sec})$. The rate coefficient at any temperature can be calculated using the Arrhenius equation $k = A \exp \{-E_a/RT\}$. Remember to use a value of R in units of $J/(K \text{ mol})$. Note: in this case the units reflect concentration expressed in molecules per cm^3.

$$k = A \exp \{-E_a/RT\}$$
$$= (9.00 \times 10^{-12} \text{ cm}^3/(\text{molecule sec})) \times \exp\{-(1.94 \times 10^3 \text{ J/mol})/$$
$$[(301K) \times (8.31 \text{J}/(\text{mol K}))]\}$$
$$= 4.14 \times 10^{-12} \text{ cm}^3/(\text{molecule sec})$$

DEMONSTRATION PROBLEM

The decomposition of $BrONO_2$ follows first-order reaction kinetics with a rate constant equal to $5.50 \times 10^{-5} \text{sec}^{-1}$: $BrONO_2 \rightarrow BrO + NO_2$. If the initial concentration of $BrONO_2$ is equal to 0.00145M, what is the concentration expected after 3.11-, 5.70-, and 217-minute time intervals?

Solution

The concentration of $BrONO_2$ as a function of time is given in the first-order integrated rate law $\ln[BrONO_2]_t = -kt + \ln[BrONO_2]_o$. Attention must be given to equating the units of time and the first-order rate constant.

Rearranging the expression to solve for $[BrONO_2]_t$:

$$\ln[BrONO_2]_t = -kt + \ln[BrONO_2]_o$$
$$[BrONO_2]_t = \exp(-kt + \ln[BrONO_2]_o)$$

Solving for time intervals of 3.11, 5.70, and 217 minutes yields:

Time (minutes)	$[N_2O_5]$
3.11	$1.44 \times 10^{-3} \text{M}$
5.70	$1.42 \times 10^{-3} \text{M}$
217	$7.09 \times 10^{-4} \text{M}$

Chapter Test

True/False

1. Most reactions exhibit faster rates at higher temperatures.

2. A first-order process with k_{rxn} equal to $1.0 \times 10^5\,sec^{-1}$ has a half-life equal to 6.9×10^{-6} seconds.

3. The overall order of any termolecular reaction is either 1 or 2.

4. The half-life of a first-order reaction depends on reactant concentration.

5. The instantaneous rate of reaction does not change with time.

Multiple Choice

6. At $500\,K$, NH_4Cl decomposes by a first-order process to form NH_3 and HCl. If the half-life of this compound is determined to be 13.0 seconds, how much time is required to decompose 91.0% of the sample?
   a. 45.2 seconds
   b. 17.7 seconds
   c. 128 seconds
   d. 1.17 seconds
   e. 75.1 seconds

7. NO_2 reacts according to the following equation: $2NO_2 \rightarrow 2NO + O_2$. What is the appearance rate of O_2 that accompanies a change in NO_2 concentration from 0.143 to $0.133\,M$ over a 33.0-second time interval?
   a. $3.03 \times 10^{-4}\,mol/(L\ sec)$
   b. $6.06 \times 10^{-4}\,mol/(L\ sec)$
   c. $1.52 \times 10^{-4}\,mol/(L\ sec)$
   d. $7.52 \times 10^{-5}\,mol/(L\ sec)$
   e. None of the above

8. Which of the conditions below are likely to *decrease* the rate of reaction?
   a. Addition of a catalyst
   b. Increase concentration of reactant
   c. Increase temperature
   d. Decrease the concentration of a reactant
   e. Addition of a reaction intermediate

9. The rate coefficient for a reaction is equal to $3.12 \times 10^{-4}\,sec^{-1}$ at $40.0°C$ and $1.04 \times 10^{-4}\,sec^{-1}$ at $20.0°C$. What is the activation energy for the reaction?
   a. $5.12 \times 10^4\,J/mol$
   b. $4.19 \times 10^4\,J/mol$
   c. $1.19 \times 10^4\,J/mol$
   d. $-7.11 \times 10^2\,J/mol$
   e. $3.20 \times 10^4\,J/mol$

10. Ozone is observed to decompose in the atmosphere according to reaction $2O_3 \rightarrow 3O_2$. What is the rate coefficient for this reaction if ozone has a half-life of 3.31 milliseconds and the decomposition follows first-order kinetic behavior?

a. $3.31 \times 10^{-3} \sec^{-1}$
b. $3.02 \times 10^{2} \sec^{-1}$
c. $2.09 \times 10^{2} \sec^{-1}$
d. $4.78 \times 10^{-3} \sec^{-1}$
e. None of the above

Short Answer

11. Write out the differential rate law for the following elementary reaction:
$NO + 2O_2 \rightarrow 2NO_2 + O_2$.

12. Ammonium nitrate decomposes by the following reaction: $NH_4NO_3 \rightarrow NH_3 + HNO_3$.
Calculate the production rate (units of mol/L sec) of NH_3 if its concentration is
observed to increase from 1.20×10^{-5} to $3.45 \times 10^{-3} M$ over a 6.01-hour period.

13. A $0.0342 M$ sample of NOBr decomposes via a first-order chemical reaction. If the
concentration of NOBr is reduced to 0.00121 after 10.1 minutes, what is the rate
coefficient for its decomposition reaction?

14. The following mechanism controls midlatitude O_3 concentrations in the lower
stratosphere:

$$O_3 + OH \rightarrow HO_2 + O_2$$

$$HO_2 + O \rightarrow OH + O_2$$

Write the net reaction for this mechanism and indicate which species acts as a
catalyst.

15. The self-reaction of NO follows a second-order rate law with k_{rxn} equal to $3.03 \times 10^{-3} L/(mol\ sec)$: $NO + NO \rightarrow N_2O_4$. If an NO sample with a initial concentration of
$0.00366 M$ is allowed to react for 17,220 seconds, what is the expected final
concentration of NO_2?

Chapter Test Answers

1. **True**

2. **True**

3. **False**

4. **False**

5. **False**

6. **a**   7. **c**   8. **d**   9. **b**   10. **c**

11. rate = $k_{rxn}[NO][O_2]^2$

12. $1.58 \times 10^{-7} mol/(L\ sec)$

13. $5.51 \times 10^{-3} \sec^{-1}$

14. $O_3 + O \rightarrow 2O_2$, catalyst: OH

15. $3.07 \times 10^{-3} M$

Check Your Performance

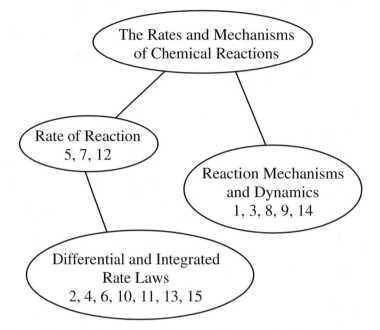

The Rates and Mechanisms
of Chemical Reactions

Rate of Reaction
5, 7, 12

Reaction Mechanisms
and Dynamics
1, 3, 8, 9, 14

Differential and Integrated
Rate Laws
2, 4, 6, 10, 11, 13, 15

Use this chart to identify weak areas, based on the question numbers you answered incorrectly in the chapter test.

Chemical Equilibrium

If given enough time, most chemical reactions stop before reaching completion. Although these systems appear static, in reality they exist in a dynamic state of chemical equilibrium. The equilibrium state is a direct consequence of the reversibility of chemical reactions. Reversible reactions consist of both forward and reverse processes that occur simultaneously in a reaction mixture. Equilibrium is a special state of a chemical system in which forward and reverse reactions occur at identical rates. Chemical equilibrium plays an enormously important role in the natural world around us. In this chapter we discuss why chemical systems react in such a manner to achieve a state of equilibrium and how equilibrium systems react to changes imposed on them. In addition, we examine several aqueous phase chemical equilibria that include acid-base reactions, solubility equilibria, and complex ion formation.

ESSENTIAL BACKGROUND

- **The quadratic equation and approximation methods**
- **Stoichiometry (Chapter 3)**
- **The rates of chemical reactions (Chapter 11)**

TOPIC 1: EQUILIBRIUM CONDITION

KEY POINTS

✓ *Why is equilibrium considered a dynamic state?*

✓ *What is the law of mass action? What is the equilibrium constant?*

✓ *What is Le Chatelier's principle? How is it used to make predictions about the direction of chemical change?*

Most chemical reactions are actually reversible processes consisting of forward and reverse reactions. A chemical system is said to be in a **condition of equilibrium** when the rate of forward reaction is exactly equal to the rate of reverse reaction. For example, the decomposition and formation of $ClONO_2$ can be written as the same chemical reaction but occurring in opposite directions: $ClONO_2 \rightleftarrows ClO + NO_2$. Thus, at equilibrium the rate of $ClONO_2$ formation exactly equals its rate of decomposition. A system at equilibrium *appears* static because there is no observable change in the concentration of reactants or products as a function of time. However, on a microscopic level, the equilibrium condition is highly dynamic. Forward and reverse chemical reactions occur continuously; however, they proceed at rates that exactly compensate for each other.

The set of concentrations of reactants and products for a system in equilibrium is called the **equilibrium position**. Experiments have shown that in many cases a great number of equilibrium positions can satisfy the condition of dynamic equilibrium. The **law of mass action** defines this set of possible equilibrium positions. This law specifies that for a given reaction of the type $aA + bB \rightleftarrows cC + dD$, the following equation characterizes all possible equilibrium positions:

$$K_c = \frac{[C]^c [D]^d}{[A]^a [B]^b}$$

In this expression, K_c is a constant called the **equilibrium constant** and the superscripts a, b, c, and d represent stoichiometric coefficients present in the balanced chemical equation.

When dealing with chemical equilibria between gaseous reactants and products, it is often more convenient to express equilibrium constants in terms of partial pressures. For example, consider the following reaction used to synthesize gaseous ammonia: $N_{2(g)} + 3H_{2(g)} \rightleftarrows 2NH_{3(g)}$. The ideal gas law predicts that gas partial pressure is proportional to concentration at a given temperature. Therefore, the equilibrium expression for this gaseous system can be written as

$$K_p = \frac{P^2_{NH_3}}{P_{N_2} P^3_{H_2}}$$

where K_p is the equilibrium constant in units of pressure and P_x represents partial pressures. For any reaction of the generic form $aA + bB \rightleftarrows cC + dD$, K_p and K_c are related by the expression $K_p = K_c (RT)^{\Delta n}$, where $\Delta n = (c + d) - (a + b)$.

Our description of equilibria has only dealt with chemical systems that exist in one phase. These equilibria are called **homogeneous equilibria**. Equilibria that involve reactants or products in more than one phase are readily observed in nature and are called **heterogeneous equilibria**. Heterogeneous equilibria are treated analogously to homogeneous equilibria; however, the equilibrium position is independent of the amounts of pure liquids and solids present. Accordingly, solids and liquids are not included in the equilibrium constant expression. For example, the equilibrium constant for the reaction $NH_4NO_{3(s)} \rightleftarrows NH_{3(g)} + HNO_{3(g)}$ is only dependent on P_{NH_3} and P_{HNO_3} and can be expressed as $K_p = P_{NH_3} P_{HNO_3}$.

Experimentation has shown that all chemical systems spontaneously react to achieve a condition of equilibrium. Therefore, knowledge of the equilibrium constant for a given chemical reaction is particularly useful for predicting the direction a reaction proceeds spontaneously. To illustrate this point, consider the reaction $OH_{(g)} + HNO_{3(g)} \rightleftarrows H_2O_{(g)} + NO_{3(g)}$, which has an equilibrium constant equal to $7.1 \times 10^{-5} \, L^2/mol^2$ at 400°C. Initially, reactants and products are mixed together with concentrations of $[HNO_3] = 0.050 \, M$, $[OH] = 0.050 \, M$, $[H_2O] = 0.0010 \, M$, and $[NO_3] = 0.0010 \, M$. The spontaneous direction of reaction for these conditions is determined by substituting initial concentrations given into the equilibrium expression to determine the **reaction quotient (Q)**:

$$Q = [H_2O][NO_3]/[OH][HNO_3]$$
$$= \{(0.0010 \, M)(0.0010 \, M)\}/\{(0.050 \, M)(0.050 \, M)\} = 4.0 \times 10^{-4}$$

Notice that the reaction quotient has an algebraic form identical to the equilibrium constant but reflects nonequilibrium conditions. A comparison of the reaction quotient and the equilibrium constant reveals what change is necessary to achieve equilibrium. If Q is greater than K, a net

Table 12.1 Rules for Predicting the Spontaneous
          Direction of Reaction

CONDITION	DIRECTION OF CHANGE
$Q > K$	Reverse reaction (products $\rightarrow$ reactants)
$Q < K$	Forward reaction (reactants $\rightarrow$ products)
$Q = K$	No spontaneous change (system is at equilibrium)

conversion of products to reactants must occur. If Q is less than K, a net conversion of reactants to products is required. In the example above, because Q is greater than K, $4.0 \times 10^{-4} > 7.1 \times 10^{-5}$, a net conversion of products to reactants is expected to occur. The guidelines for predicting the spontaneous direction of a chemical reaction are summarized in **Table 12.1**.

In the example above, net conversion of products to reactants will eventually lead to a state of dynamic equilibrium. The reactant and product concentrations that satisfy the equilibrium condition can also be calculated from the information given above. In this calculation, initial and equilibrium concentrations of reactants and products are connected by a variable (x) indicating the extent of reaction:

	$OH_{(g)}$	+	$HNO_{3(g)}$	$\rightleftarrows$	$H_2O_{(g)}$	+	$NO_{3(g)}$
Initial concentration (M)	0.050		0.050		0.0010		0.0010
Change	$+x$		$+x$		$-x$		$-x$
Equilibrium	$0.050 + x$		$0.050 + x$		$0.0010 - x$		$0.0010 - x$

The equilibrium position is determined by substituting the equilibrium concentrations, as a function of x, into the definition of the equilibrium constant and solving:

Substituting equilibrium concentrations into the equilibrium expression:

$$K_c = 7.1 \times 10^{-5}\, L^2/mol^2 = [H_2O][NO_3]/[OH][HNO_3] = (0.0010 - x)^2 \big/ (0.050 + x)^2$$

Expanding and combining like terms:

$$7.1 \times 10^{-3} = (4 \times 10^{-4} - 0.04x + x^2)\big/(1 \times 10^{-4} + 0.02x + x^2) - 0.993x^2 + 0.0201x - 9.93 \times 10^{-5} = 0$$

Solving via the quadratic equation:

$$x = \frac{-b \pm \sqrt{b^2 - 4ac}}{2a} = \frac{2.007 \times 10^{-3} \pm \sqrt{4.028 \times 10^{-6} - 4(0.9998)(8.225 \times 10^{-7})}}{2(0.9998)}$$

$$x = 5.8 \times 10^{-4} \quad \text{and/or} \quad 1.4 \times 10^{-3}$$

Note: The value of x cannot be 1.4×10^{-3} because substitution into the equilibrium concentration expressions yields negative values for H_2O and NO_3.

Solving for equilibrium position:

$$[H_2O] = [NO_3] = (0.0010\,M - 5.8 \times 10^{-4}\,M) = 4.2 \times 10^{-4}\,M$$

$$[OH] = [HNO_3] = (0.05\,M + 5.8 \times 10^{-4}\,M) = 5.1 \times 10^{-2}\,M$$

A sequence of procedures commonly used to solve equilibrium position problems is summarized in **Table 12.2**. The determination of equilibrium concentrations using this method is often simplified considerably by using approximation methods (see Demonstration Problem).

Table 12.2 Strategy for Solving Equilibrium Position Problems

1. Write out a balanced equation for the chemical system.
2. Calculate Q for initial conditions and determine the direction of shift to equilibrium (Table 12.1).
3. Construct a chart defining initial conditions, change for equilibrium, and equilibrium concentrations:

	A	$+$	B	$\rightleftarrows$	C
Initial	1M		1M		0
Change:	$-x$		$-x$		x
Equilibrium	$1M - x$		$1M - x$		x

4. Substitute equilibrium concentrations (in terms of x) into the equilibrium expression:

$$K_c = \frac{x}{[1M - x][1M - x]}$$

5. Solve for equilibrium concentrations using *basic algebra*, the *quadratic equation*, or *approximation methods*.
6. Substitute equilibrium concentrations into expression for K_c to check calculation (optional).

Table 12.3 Le Chatelier's Principle as Applied to the Reaction $N_{2(g)} + 3H_{2(g)} \rightleftarrows 2NH_{3(g)}$ $\Delta H° = -45.9\,kJ$

STRESS APPLIED	AFFECT ON EQUILIBRIUM POSITION
$\uparrow [NH_3]$	Shift to the *left*: $N_{2(g)} + 3H_{2(g)} \leftarrow 2NH_{3(g)}$
$\downarrow [NH_3]$	Shift to the *right*: $N_{2(g)} + 3H_{2(g)} \rightarrow 2NH_{3(g)}$
$\uparrow [H_2]$ or $\uparrow [N_2]$	Shift to the *right*: $N_{2(g)} + 3H_{2(g)} \rightarrow 2NH_{3(g)}$
$\downarrow [H_2]$ or $\downarrow [N_2]$	Shift to the *left*: $N_{2(g)} + 3H_{2(g)} \leftarrow 2NH_{3(g)}$
$\uparrow$ Volume	Shift to the *left*: $N_{2(g)} + 3H_{2(g)} \leftarrow 2NH_{3(g)}$
$\downarrow$ Volume	Shift to the *right*: $N_{2(g)} + 3H_{2(g)} \rightarrow 2NH_{3(g)}$
$\uparrow$ Temperature	Shift to the *left*: $N_{2(g)} + 3H_{2(g)} \leftarrow 2NH_{3(g)}$
$\downarrow$ Temperature	Shift to the *right*: $N_{2(g)} + 3H_{2(g)} \rightarrow 2NH_{3(g)}$

A system at equilibrium is in a dynamic state of balance. Therefore, stress placed on the system often disturbs this balance and moves the system out of equilibrium. The effect of a stress imposed on an equilibrium system can be qualitatively predicted using **Le Chatelier's principle**, which states *if a system at equilibrium is subjected to a change in concentration, pressure, or temperature, it will respond in a manner to counteract the effect of the disturbance.* To illustrate the significance of this principle, consider the following system at equilibrium: $N_{2(g)} + 3H_{2(g)} \rightleftarrows 2NH_{3(g)}$ $\Delta H° = -45.9\,kJ$. From Le Chatelier's principle, it follows that the addition of a reactant or product initiates a reaction that consumes the added material. Conversely, if a substance is removed, the system will react in such a way to reform the removed material. Le Chatelier's principle also predicts that a reduction in volume available to a gaseous system at equilibrium will initiate a chemical reaction that decreases the total number of moles of gas present. Thus, it follows in the example above that decreasing the reactor volume initiates a net conversion of N_2 and H_2 to NH_3. Finally, Le Chatelier's principle predicts that when temperature is raised, an equilibrium system will react in a manner to absorb the heat provided. Because endothermic reactions consume energy, it follows that an increase in temperature initiates a net increase in the extent of the endothermic direction of a chemical system in equilibrium. Therefore, in the example above, increasing temperature results in conversion of NH_3 to N_2 and H_2. Conversely, if temperature is decreased, an increase in the extent of the exothermic direction is predicted. To reinforce the use of Le Chatelier's principle, predictions made by applying this principle to the N_2, H_2, and NH_3 systems are summarized in **Table 12.3**.

Topic Test 1: Equilibrium Condition

True/False

1. A system at equilibrium is characterized by a state of no chemical change.

2. When a stress is placed on a system at equilibrium it will respond in a way to minimize the disturbance.

3. For a system at equilibrium, the rate of forward reaction equals the rate of reverse reaction.

Multiple Choice

4. The concentrations of a gaseous mixture at equilibrium are $[NO] = 0.320\,M$, $[O_2] = 0.545$, and $[NO_2] = 0.00321\,M$. Calculate the equilibrium constant for the system $2NO_{(g)} + O_{2(g)} \rightleftarrows 2NO_{2(g)}$.
   a. $1.85 \times 10^{-4}\,L/mol$
   b. $1.84 \times 10^{-2}\,L/mol$
   c. $5.91 \times 10^{-5}\,L/mol$
   d. $5.75 \times 10^{-5}\,L/mol$
   e. None of the above

5. The following chemical system has an equilibrium constant (K_p) equal to $1.01\,atm$ at 750°C: $CaCO_{3(s)} \rightleftarrows CaO_{(s)} + CO_{2(g)}$. What is the equilibrium pressure of CO_2 expected when a 4.50×10^5 gram sample of $CaCO_3$ is in equilibrium with its reaction products?
   a. $2.24 \times 10^{-6}\,atm$
   b. $4.54 \times 10^5\,atm$
   c. $4.50 \times 10^{-5}\,atm$
   d. $1.01\,atm$
   e. $4.50\,atm$

Short Answer

6. Given the thermodynamic equation $NO_{2(g)} + NO_{3(g)} \rightleftarrows N_2O_{5(g)}$ $\Delta H = -54.0\,kJ/mol$, predict the response of the system to the following physical changes: (1) addition of NO_2, (2) increase in temperature, (3) increase in volume, and (4) removal of N_2O_5.

7. A 1.000-atm sample of NO_2 is placed in an evacuated cylinder and allowed to reach equilibrium: $NO_{2(g)} + NO_{2(g)} \rightleftarrows N_2O_{4(g)}$. If the partial pressure of N_2O_4 is equal to $0.478\,atm$ at equilibrium, what is the equilibrium constant (K_p) for the reaction?

Topic Test 1: Answers

1. **False.** A system at equilibrium is dynamic. Forward and reverse reactions occur at the same rate.

2. **True.** This is a restatement of Le Chatelier's principle.

3. **True.** The equilibrium condition is characterized by equal rates of forward and reverse reactions.

4. **a.** The law of mass balance defines the equilibrium constant in the following expression: $K_c = [NO_2]^2/([NO^2] [O_2])$. The value of K_c is determined by substituting equilibrium concentrations into this equation:

$$K_c = (0.00321 M)^2 \Big/ \left[(0.320 M)^2 (0.545 M) \right] = \textbf{1.85} \times \textbf{10}^{-4} \, \textbf{L/mol}$$

5. **d.** Solids are excluded from the equilibrium constant expression in heterogeneous equilibria. Therefore, at equilibrium $K_p = P_{CO_2} = 1.01$ atm.

6. Forward, reverse, reverse, forward. (1) The system will react to the addition of NO_2 by consuming the added substance. This proceeds via the forward direction. (2) Increasing the temperature will result in a net increase in the endothermic reaction. In this example, the reverse reaction is endothermic. (3) Increasing the volume will result in a net production of moles of gas. Thus, the reverse reaction that makes 2 moles of gas will be favored. (4) The system will react to reform the N_2O_5 removed. Thus, the forward reaction will occur.

7. $247 \, \text{atm}^{-1}$. The equilibrium constant is determined by substituting in the equilibrium partial pressures to the expression $K_p = P_{N_2O_2}/P_{NO_2}^2$. Remember, the formation of $1 N_2O_2$ molecule consumes $2 NO_2$ molecules.

   Calculating the equilibrium partial pressure of NO_2:

   $$[NO_2]_{eq} = [NO_2]_{initial} - 2[N_2O_4]_{eq} = 1.00 - 2(0.478) = 0.044 \, \text{atm}$$

   Calculating the equilibrium constant:

   $$K_p = P_{N_2O_2} \Big/ P_{NO_2}^2 = (0.478 \, \text{atm})/(0.044 \, \text{atm})^2 = \textbf{247 atm}^{-1}$$

TOPIC 2: ACID-BASE EQUILIBRIA

KEY POINTS

✓ *What are strong and weak acids? What are strong and weak bases?*

✓ *How is acid or base concentration determined by titration?*

✓ *What are the characteristics of a buffered system?*

Many of the materials we deal with on a daily basis possess acidic or basic properties. Acids and bases can be loosely defined in terms of conjugate pairs in the **Bronsted-Lowry model**. In this description, an acid is a substance that can donate a proton and a base is a substance that accepts a proton. All acids and bases are also characterized in terms of how easily they are ionized in solution. **Strong acids** are compounds that completely dissolve in solution to generate H^+ ions. An example of a strong acid is HNO_3, which dissociates in solution according to the equation $HNO_{3(l)} \rightarrow H^+_{(aq)} + NO^-_{3(aq)}$. In contrast, **weak acids** only partially dissociate in solution to generate H^+ cations. Therefore, weak acids exist in both ionized and non-ionized forms in solution. Accordingly, the generic reaction $HA_{(aq)} + H_2O_{(l)} \rightleftharpoons A^-_{(aq)} + H_3O^+_{(aq)}$ represents the dissolution of a weak acid in H_2O. As shown in **Figure 12.1**, **HA/A⁻** and **H₃O⁺/H₂O** can be regarded as **conjugate acid** and **base pairs** because they differ only in the presence of a transferred proton. The equilibrium expression for the dissociation of an acid in water is expressed in the following equation:

$$K_a = \frac{[H_3O^+][A^-]}{[HA]} = \frac{[H^+][A^-]}{[HA]}$$

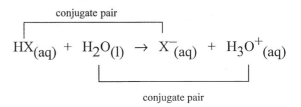

$$\overbrace{\text{HX}_{(aq)} + \text{H}_2\text{O}_{(l)} \rightarrow \underbrace{\text{X}^-_{(aq)} + \text{H}_3\text{O}^+_{(aq)}}}$$

conjugate pair

conjugate pair

Figure 12.1. Conjugate acid-base pairs for dissolution of an acid in water.

where H_3O^+ or H^+ are used interchangeably and represent the transferred proton. K_a is the equilibrium constant in this equation and is called the **acid dissociation constant**. The magnitude of K_a relates to how easily a hydrogen ion is transferred in solution and is proportional to an acid's strength. A large value of K_a indicates an acid that dissociates appreciably when dissolved in water. **Polyprotic acids** are compounds that have more than one ionizable proton and, therefore, have several values of K_a corresponding to the loss of successive H^+ ions.

Strong bases are compounds that completely dissolve in solution to produce OH^- anions. An example of a strong base is NaOH, which dissociates in solution according to the following equation: $\text{NaOH}_{(s)} \rightarrow \text{Na}^+_{(aq)} + \text{OH}^-_{(aq)}$. In contrast, **weak bases** exist in partially ionized forms in water. The dissolution of a base in solution can be generically written as $\text{B}_{(aq)} + \text{H}_2\text{O}_{(l)} \rightleftharpoons \text{BH}^+_{(aq)} + \text{OH}^-_{(aq)}$, where B/BH^+ and $\text{H}_2\text{O}/\text{OH}^-$ are conjugate pairs. The equilibrium constant for this reaction is called the **base-dissociation constant** and is defined by the equation:

$$K_b = \frac{[\text{OH}^-][\text{HB}^+]}{[\text{B}]}$$

The acid dissociation and base dissociation constants for conjugate acid-base pairs are inversely proportional to each other. This relationship is quantitatively expressed by the equation $\mathbf{K_w = K_a K_b}$, where K_w is the **ion-product constant** for water auto-ionization: $\text{H}_2\text{O} \rightleftharpoons \text{H}^+_{(aq)} + \text{OH}^-_{(aq)}$, $K_w = 1.00 \times 10^{-14} \text{M}^2$ at 25°C. For example, the dissolution of HF into water establishes the following equilibrium between conjugate acid-base pairs: $\text{HF}_{(aq)} \rightleftharpoons \text{H}^+_{(aq)} + \text{F}^-_{(aq)}$. Given that K_a is equal to $7.2 \times 10^{-4} \text{M}$, the value of K_b for F^- is calculated by rearranging the expression $K_w = K_a \times K_b$ and solving for K_b:

$$K_b = K_w/K_a = (1.0 \times 10^{-14} \text{M}^2)/(7.2 \times 10^{-4} \text{M}) = 1.4 \times 10^{-11} \text{M}$$

The concentrations of conjugate acid-base pairs in solution can be calculated given the appropriate acid-base dissociation constant and initial concentrations. For example, the concentration of H^+ and CN^- in a 2.5 M solution of HCN ($K_a = 6.3 \times 10^{-10}$) can be determined by the following equilibrium calculation.

	$\text{HCN}_{(aq)}$	$\rightleftharpoons$	$\text{H}^+_{(aq)}$	+	$\text{CN}^-_{(aq)}$
Initial concentration (M)	2.5		≈ 0		0
Change	$-x$		$+x$		$+x$
Equilibrium	$2.5 - x$		x		x

Substituting equilibrium concentrations into the equilibrium expression:

$$K_c = 6.3 \times 10^{-10} = [\text{H}^+][\text{CN}^-]/[\text{HCN}] = x^2/(2.5 - x)$$

The quadratic expression is easily solved via approximation (see Demonstration Problem). The extent of dissociation (x) is expected to be small because the value of K_a is very small. Therefore, we may assume that in the denominator $(2.5 - x) \cong 2.5$.

Solving via approximation:

$$6.3 \times 10^{-10} = x^2/2.5$$

$$x = 4.0 \times 10^{-5}$$

Determining the equilibrium position:

$$[H^+] = [CN^-] = x = 4.0 \times 10^{-5}\,M$$

$$[HCN] = (2.5 - x) = 2.5 - 4.0 \times 10^{-5} \cong 2.5\,M$$

The use of approximation methods is extremely useful when solving for weak acid and base equilibrium positions and is generally applicable when the value of K_a is small.

The extent that a weak acid or weak base is dissociated in solution is strongly affected by the presence of a **common ion**. In this context, a common ion is a species that is present in the equilibrium expression of a weak acid but is added to solution independently. For example, consider the dissociation of HCN to produce H^+ and CN^- treated above. According to Le Chatelier's principle, the addition of a soluble compound that can produce CN^- ions, such as $NaCN_{(s)}$, causes a shift in the equilibrium position in the direction that consumes CN^-. Therefore, the presence of additional CN^- ions causes the reformation of HCN and decreases the net amount of acid dissociated. Calculations of acid-base equilibrium positions always need to include the affect of any common ions present.

As discussed in Chapter 3, strong acids and bases react to completion via neutralization: $HA_{(aq)} + OH^-_{(aq)} \rightarrow H_2O_{(l)}$. Such neutralization reactions are often used to determine the concentrations of acids and bases in solution via a **titration analysis**. In an acid-base titration, a solution containing a base of known concentration is added to an acidic solution of an unknown concentration. Acid-base indicators are often used to mark the **equivalence point**, which is defined as the point at which an equivalent amount of base has been added to completely neutralize the unknown acid solution. In addition to acid-base indicators, a solution's pH can be monitored to determine the equivalence point of a titration. **pH** is a convenient way of expressing the concentration of H^+ in solution and is mathematically expressed as $\mathbf{pH = -\log([H^+])}$ or equivalently $\mathbf{[H^+] = 10^{-pH}}$. Thus, acidic solutions have low values of pH and basic solutions have high values of pH. **Figure 12.2** shows a plot of pH vs. volume of base added for the titration of 10 mL HCl solution with 0.01 M NaOH. As shown in the titration curve, the equivalence point in a strong acid-strong base titration is marked by the point of inflection in the plot and occurs at a pH of 7. The number of moles of OH^- added at the equivalence point is exactly equal to the number of moles of H^+ present in the unknown solution. This value can be used to determine the unknown acid concentration by dividing by the original volume of acid. In the example in Figure 12.2, this corresponds to a concentration of 0.05 M HCl:

$$[HCl] = [(0.050\,L) \times (0.01\,M)]/0.010\,L = \mathbf{0.05\,M\ HCl}$$

Analogously, the concentration of a strong base can be determined by titration with a strong acid.

Weak acids and bases also react via neutralization when titrated with a strong base or acid but do not react to completion. Figure 12.2 shows the titration curve of a 10-mL sample of HOCl ($K_a = 3.5 \times 10^{-8}$) with a 0.01 M NaOH solution. As shown in Figure 12.2, the pH change in the region of the equivalence point is significantly more gradual than that observed in the strong acid-base titration. In addition, the equivalence point does not occur at a pH of 7. The differences in the titration curves of weak acids and bases are related to the formation of buffered

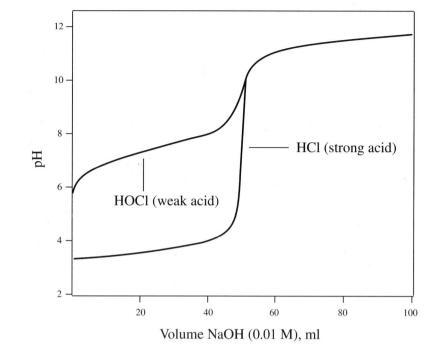

Figure 12.2. Titration curves for 10-mL samples of 0.05 M HCl and 0.05 M HOCl ($K_a = 3.5 \times 10^{-8}$) with 0.01 M NaOH.

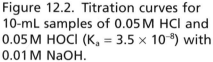

solutions. **Buffered solutions** are solutions that resist a change in pH upon addition of a strong acid or base. These solutions are composed of a mixture of a weak acid and a weak base that are commonly conjugate acid and base pairs. Because they contain both an acid and base, the addition of common ions such as H^+ or OH^- merely acts to shift the equilibrium abundance of conjugate acid and base pairs in accordance to Le Chatelier's principle. The scavenging of H^+ and OH^- results in a significantly reduced change in pH upon addition of a strong acid or strong base. The pH of a buffer system is commonly calculated using the **Henderson-Hasselbalch equation**:

$$pH = pK_a + \log\left(\frac{[A^-]}{[HA]}\right)$$

where $[A^-]$ and $[HA]$ are the equilibrium concentrations of the conjugate base and conjugate acid and pK_a is the negative base 10 logarithm of the value of K_a: $\mathbf{pK_a = -\log(K_a)}$ or equivalently $\mathbf{K_a = 10^{-pKa}}$. Usually, the amounts of acid and base that ionize in solution are small enough so that the initial acid and base concentrations can be used directly in the Henderson-Hasselbalch equation.

Topic Test 2: Acid-Base Equilibria

True/False

1. A strong base dissociates completely in solution.

2. A buffered solution contains a strong acid and its conjugate base.

3. The equivalence point for the titration of a weak base with a strong acid occurs at a pH of 7.

Multiple Choice

4. A 51.0-mL unknown solution of HCl is titrated with a 0.0311 M KOH solution. If the equivalence point is reached after the addition of 15.5 mL base, what is the concentration of the unknown acid?
   a. 3.11×10^{-2} M
   b. 1.84×10^{-2} M
   c. 1.06×10^2 M
   d. 4.82×10^{-4} M
   e. 9.45×10^{-3} M

5. HOCl is a weak acid with a K_a equal to 3.6×10^{-8} M at 28°C. What is the pH of a 0.100 M solution of HOCl?
   a. 4.2
   b. 7.0
   c. −1
   d. 13
   e. 5.5

Short Answer

6. The K_a for HF is equal to 7.2×10^{-4} M. Calculate the pH of a buffer solution made by adding 2.2 liters of a 0.145 M HF solution to 3.1 liters of a 0.154 M NaF solution. Assume NaF completely dissociates via the equation $NaF_{(s)} \rightarrow F^-_{(aq)} + Na^+_{(aq)}$.

Topic Test 2: Answers

1. **True.** Strong bases dissociate completely to generate OH^- anions in solution.

2. **False.** A buffer is a solution composed of a mixture of a weak acid and a weak base.

3. **False.** The equivalence point for the titration of a weak base with a strong acid occurs at a pH < 7.

4. **e.** The concentration of the unknown HCl sample is obtained by dividing the number of moles delivered at the equivalence point by the initial volume of the sample.

   [HCl] = [(0.0311M)(0.0155L)]/(0.0510 liters) = **9.45×10^{-3} M**

5. **a.** The [H^+] at equilibrium is calculated using the initial acid concentration and the acid-dissociation constant. Once the [H^+] is calculated, the pH is determined using the equation pH = −log([H^+]).

   Solving for $[H^+]_{eq}$:

	$HOCl_{(aq)}$	$\rightleftharpoons$	$H^+_{(aq)}$	+	$OCl^-_{(aq)}$
Initial concentration (M)	0.100		≈0		0
Change	−x		+x		+x
Equilibrium	0.100 − x		x		x

   The equilibrium position can be solved via approximation (see Demonstration Problem) because K_a is very small. $x^2/0.10 \approx 3.6 \times 10^{-8}$ M, therefore, $x = [H^+] = 6.0 \times 10^{-5}$ M.

Calculating the solution pH:

$$pH = -\log([H^+]) = -\log(6.0 \times 10^{-5}\,M) = \textbf{4.2}$$

6. 3.3. The pH of the resulting buffer solution can be calculated using the Henderson-Hasselbalch equation (pH = pK$_a$ + log([A$^-$]/[HA]). Remember to include the effect of dilution in your calculation of HF and F$^-$ concentrations:

Solving for pH via Henderson-Hasselbalch equation:

$$pH = pKa + \log([F^-]/[HF]) = 3.14 + \log[(0.090)/(0.060)] = \textbf{3.3}$$

TOPIC 3: SOLUBILITY AND COMPLEX ION FORMATION

KEY POINTS

✓ *What is the solubility product constant?*

✓ *Which factors influence a compound's solubility?*

✓ *What is a complex ion? How is it formed in solution?*

The dissolution of a soluble compound in water is another process that involves the principles of dynamic equilibrium. For example, when an excess of solid barium fluoride is added to a nearly saturated solution, the following equilibrium is rapidly established between undissolved BaF$_2$ and its hydrated ions: BaF$_{2(s)}$ ⇌ Ba$^{2+}_{(aq)}$ + 2F$^-_{(aq)}$. The equilibrium constant for this process is called the **solubility product (K$_{sp}$)** and depends only on the concentration of ions present at equilibrium. For example, the equilibrium constant expression for the solvation of BaF$_2$ is written as K$_{sp}$ = [Ba^{2+}] [F$^-$]^2. Like most equilibrium constants, K$_{sp}$ is independent of most environmental factors with the exception of temperature. The **solubility** of a compound is the amount that dissolves in a pure solvent to form a saturated solution and is often reported in units of grams per liter. Similarly, **molar solubility** is the number of moles that dissolve to form a liter of saturated solution. The solubility product and molar solubility are closely related to each other via the value of K$_{sp}$ and the stoichiometry of the balanced chemical equation. For example, given that K$_{sp}$ for BaF$_2$ is equal to 1.00×10^{-6}, the molar solubility of BaF$_2$ in pure water can be calculated in the following manner:

	BaF$_{2(s)}$	⇌	Ba$^{2+}_{(aq)}$	+	2F$^-_{(aq)}$
Initial concentration (M)	——		0		0
Change	——		+x		+2x
Equilibrium	——		x		2x

Substituting equilibrium concentrations into the equilibrium constant expression:

$$K_{sp} = 1.00 \times 10^{-6}\,M = [Ba^{2+}][F^-]^2 = (x)(2x)^2 = 4x^3$$

Solving for x:

$$x = \textbf{6.33} \times \textbf{10}^{-3}\,\textbf{M}$$

Therefore, in 1 liter of solution, 6.33×10^{-3} moles of BaF$_2$ is expected to dissolve to form a saturated solution.

Unlike the solubility product, the solution-phase environment can dramatically affect the molar solubility of a compound. Specifically, molar solubility differs in impure water conditions because of the presence of a common ion, the solution pH, and the presence of complexing agents. The direction in which these factors affect solubility may be predicted by applying Le Chatelier's principle in each case. The presence of a common ion in solution always acts to decrease the observed molar solubility. This behavior is expected from Le Chatelier's principle because the addition of a common ion is a stress that results in a net conversion from product ions to solid reactant.

The solubilities of ionic compounds that dissolve to produce either acidic or basic ions are affected by solution pH. For example, consider the dissolution of $Ni(OH)_2$ into pure water: $Ni(OH)_{2(s)} \rightleftarrows Ni^{2+}_{(aq)} + 2OH^-_{(aq)}$, $K_{sp} = 2.0 \times 10^{-15} M^2$. The solubility in pure water is predicted to be $8.9 \times 10^{-6} M$. This corresponds to a slightly basic solution of pH 8.9. How will the solubility be affected by lowering the solution pH? The solubility of $Ni(OH)_2$ in a solution buffered to pH of 7.0 can be calculated by substituting the [OH] expected at this pH into the equilibrium expression ($[OH] = 1.0 \times 10^{-7} M$):

$$K_{sp} = [Ni^{2+}][OH^-]^2 = [Ni^{2+}][1.0 \times 10^{-7} M]^2 = 2.0 \times 10^{-15} M^2$$

Solving for $[Ni^{2+}]$:

$$[Ni^{2+}] = 0.20 M \text{ (a factor 22,000 times greater)}$$

Thus, the solubility of slightly soluble ionic compounds containing a basic ion increases as pH is lowered. Similarly, the solubility of ionic salts containing acidic ions increases as pH is raised.

In addition to forming solids in solution, ions can associate to form **complex ions**. Formation of a complex ion involves the attraction of electrons from an electron pair donor called a **Lewis base** by an electron pair acceptor called a **Lewis acid**. Typically, this interaction results in the formation of a covalent bond between Lewis acids and bases. Complex ions are often formed by the interaction of a metal ion with one or more **ligands**. In these cases, the metal ion acts as an electron pair acceptor and the ligands are species that can share lone pair electrons. Complex ions are characterized by **coordination number**, which is the number of ligands that are covalently bound to a single metal ion. Complex ion formation also involves a dynamic equilibrium between ionic and complex ionic forms. For example, the formation of $Ag(CN)_2^-$ proceeds through the following complex ion reaction: $Ag^+_{(aq)} + 2CN^-_{(aq)} \rightleftarrows Ag(CN)^-_{2(aq)}$, $K_f = 1 \times 10^{24} L/mol$, where K_f is called the **formation constant**. The values of K_f vary from system to system and can be very large. Therefore, addition of the appropriate ligand can dramatically increase the solubility of metal-containing salts.

Topic Test 3: Solubility and Complex Ion Formation

True/False

1. Formation of a complex ion reduces the solubility of a metal-containing salt.

2. The solubility of some solids depends on pH.

3. The value of K_{sp} depends on the concentration of common ion present.

Multiple Choice

4. Iron (II) hydroxide $[Fe(OH)_2]$ is a slightly soluble metal salt that has a K_{sp} equal to $7.00 \times 10^{-16} M^2$ at 25°C. What is the molar solubility of $Fe(OH)_2$ in pure water at 25°C?

a. 2.65×10^{-8} M
b. 5.60×10^{-6} M
c. 1.32×10^{-8} M
d. 3.5×10^{-16} M
e. None of the above

5. Which of the following salts is expected to be most soluble in a basic solution?
   a. NaOH
   b. NaCl
   c. NH_4Cl
   d. $Ca(OH)_2$
   e. KF

Short Answer

6. Calculate the solubility of $AgCl_{(s)}$ ($K_{sp} = 1.9 \times 10^{-10}$) in a 0.100 M NaCl solution. Assume complete dissociation of NaCl in solution.

Topic Test 3: Answers

1. **False.** Complex ion formation increases solubility because it removes metal ions from solution.

2. **True.** The solubility of salts that contain acidic or basic ions is expected to depend on pH.

3. **False.** K_{sp} is independent of the solution phase composition.

4. **b.** Given K_{sp}, the molar solubility can be deduced by the following calculation:

   $$K_{sp} = 7.0 \times 10^{-16} \, M^3 = [Fe^{2+}][OH^-]^2 = x(2x)^2 = 4x^3$$

   Solving: $x = \textbf{molar solubility} = \textbf{5.60} \times \textbf{10}^{-6}$ **M**

5. **c.** NH_4^+ reacts with the excess OH^- present in a basic solution via the equation: $NH_{4(aq)}^+ + OH_{(aq)}^- \rightarrow NH_{3(aq)} + H_2O_{(l)}$. This reaction consumes NH_4^+ and, thus, enhances the solubility of NH_4Cl.

6. 1.9×19^{-9} M. The presence of a common ion reduces the solubility of AgCl. Solubility under these conditions can be calculated by

   $$K_{sp} = 1.9 \times 10^{-10} \, M^2 = [Ag^+][Cl^-] = (x)(0.100 + x)$$

   Using approximate methods:

   $$K_{sp} = 1.9 \times 10^{-10} \, M^2 = (x)(0.10), \text{ solving yields } x = \textbf{molar solubility} = \textbf{1.9} \times \textbf{10}^{-9} \, \textbf{M}$$

DEMONSTRATION PROBLEM

Formic acid (HCOOH) is an important trace acid in the chemistry of marine rain water. Given that formic acid has a K_a equal to 1.80×10^{-4} M at 25°C, calculate the concentrations of HCOOH, H^+, and $HCOO^-$ in a 0.540 M solution at 25°C. Use the *quadratic equation* and *approximation methods*.

Solution

HCOOH dissociates in solution to establish an equilibrium with its conjugate base: $HCOOH_{(aq)}$ $\rightleftarrows HCOO^-_{(aq)} + H^+_{(aq)}$. The equilibrium concentrations of each species can be calculated given an initial $[HCOOH] = 0.540\,M$.

	$HCOOH_{(aq)}$ $\rightleftarrows$	$HCOO^-_{(aq)}$ +	$H^+_{(aq)}$
Initial concentration (M)	0.540	0	≈0
Change	$-x$	$+x$	$+x$
Equilibrium (M)	$0.540 - x$	x	x

The equilibrium position can be solved via approximation because K_a is very small. In this approximation, we assume that $(0.540 - x) \cong 0.540$ because the value of x is likely to be very small relative to 0.540. Therefore, substitution into the equilibrium expression yields

$$K_a = 1.80 \times 10^{-4}\,M = ([H^+][HCOO^-])/[HCOOH] \approx x^2/0.540\,M$$

Solving for x:

$$x^2 = (0.540)(1.80 \times 10^{-4}) = 9.86 \times 10^{-3}\,M$$

Substituting into expressions for equilibrium concentrations:

$$x = [H^+] = [HCOO^-] = 9.86 \times 10^{-3}\,M,\ [HCOOH] = 0.540 - x = 0.530\,M$$

This problem can also be solved exactly using the quadratic equation:

$$K_a = 1.80 \times 10^{-4}\,M = ([H^+][HCOO^-])/[HCOOH] = x^2/(0.540 - x)$$

$$x^2 + 1.80 \times 10^{-4}\,x - 9.72 \times 10^{-5} = 0,$$

$$x = \frac{-1.80 \times 10^{-4} \pm \left[(1.80 \times 10^{-4})^2 - 4(9.72 \times 10^{-5})\right]^{1/2}}{2} = 9.77 \times 10^{-3}\,M$$

Chapter Test

True/False

1. At equilibrium, the rates of forward and reverse reaction are equal to 0.

2. The pH of a 0.1 M solution of HCl (strong acid) is equal to 1.0.

3. NO_2^- is the conjugate base of HNO_2.

4. Complex ion formation tends to decrease solubility.

5. Ligands act as electron pair acceptors in most complex ion formation reactions.

6. The equivalence point occurs at a pH of 7 in a titration of a strong base with a strong acid.

Multiple Choice

7. K_c for the reaction $N_{2(g)} + 3H_{2(g)} \rightleftarrows 2NH_{3(g)}$ is equal to $5.95 \times 10^{-2}\,L^2/mol^2$ at 227°C. What is the value of K_p for this process at 227°C?

a. $1.45 \times 10^{-3}\,\text{atm}^{-2}$
b. $3.53 \times 10^{-5}\,\text{atm}^{-2}$
c. $2.44\,\text{atm}^{-2}$
d. $100\,\text{atm}^{-2}$
e. None of the above

8. What is the pH of a 0.300 M solution of methylamine (weak base, CH_3NH_2: $K_b = 4.38 \times 10^{-4}\,M$) at 25°C?
   a. 10.1
   b. 3.45
   c. 1.93
   d. 12.1
   e. 13.5

9. How will the following system at equilibrium respond to an increase in temperature?
   $2N_2O_{5(g)} \rightleftarrows 4NO_{2(g)} + O_{2(g)}$ $\Delta H = 110.02\,\text{kJ}$
   a. Decrease partial pressure of O_2
   b. Increase partial pressure of N_2O_5
   c. Increase partial pressure of NO_2
   d. Decrease partial pressure of NO_2
   e. No effect

10. What is the molar solubility of CaF_2 ($K_{sp} = 3.4 \times 10^{-11}$) in a 0.0500 M Ca^{2+} solution?
    $CaF_{2(s)} \rightleftarrows Ca^{2+}_{(aq)} + 2F^-_{(aq)}$
    a. $1.3 \times 10^{-5}\,M$
    b. $1.7 \times 10^{-10}\,M$
    c. $2.1 \times 10^{-4}\,M$
    d. $2.6 \times 10^{-4}\,M$
    e. None of the above

11. Calculate the pH of a buffer solution made by combining 2.00 liters of 0.0132 M NaF solution and 3.00 liters of 0.00851 M HF ($K_a = 7.21 \times 10^{-4}\,M$) solution? Assume that NaF dissociates completely in solution.
    a. 3.32
    b. 3.16
    c. 2.96
    d. 7.00
    e. 1.80

Short Answer

12. In the titration of 72.1 milliliters of an unknown KOH solution, 21.1 milliliters of a 0.451 M HNO_3 solution are required to reach the equivalence point. What is the concentration of the unknown solution?

13. Ag^+ reacts with NH_3 to form a complex ion according to the following net chemical equation: $Ag^+_{(aq)} + 2NH_3 \rightleftarrows Ag(NH_3)^+_{2(aq)}$ $K_F = 1.7 \times 10^7$. If a solution is made by adding 500 milliliters of a 2.0 M NH_3 to 500 milliliters of a 0.20 M $AgNO_3$ solution, how much Ag^+ is present at equilibrium?

14. Identify the conjugate acid and base pairs involved with each acid-base reaction.
    a. $HF_{(aq)} + H_2O \rightleftarrows H_3O^+_{(aq)} + F^-_{(aq)}$
    b. $NH_{3(aq)} + NH_3 \rightleftarrows NH^+_{4(aq)} + NH^-_{2(aq)}$

15. An enormous amount of the Earth's carbon is cycled through ocean sediment in the form of $CaCO_{3(s)}$. If the K_{sp} for $CaCO_3$ is $4.80 \times 10^{-9}\,M^2$, what is the molar solubility of $CaCO_3$? $CaCO_{3(s)} \rightleftarrows Ca^{2+}_{(aq)} + CO^{2-}_{3(aq)}$?

16. The pH of human arterial blood is 7.34. What is the hydrogen ion (H^+) and hydroxide ion (OH^-) concentration in arterial blood?

Chapter Test Answers

1. **False**

2. **True**

3. **True**

4. **False**

5. **False**

6. **True**

7. **b**   8. **a**   9. **c**   10. **a**   11. **b**

12. $0.132\,M$

13. $7.4 \times 10^{-9}\,M$

14. a. HF/F^-, H_3O^+/H_2O; b. NH^+_4/NH_3, NH_3/NH^-_2

15. $6.93 \times 10^{-5}\,M$

16. $[H^+] = 4.6 \times 10^{-8}\,M$, $[OH^-] = 2.2 \times 10^{-7}\,M$

Check Your Performance

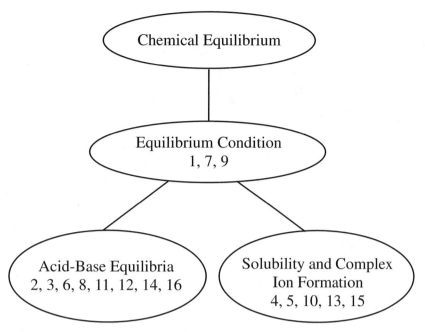

Use this chart to identify weak areas, based on the question numbers you answered incorrectly in the chapter test.

CHAPTER 13

Entropy, Free Energy, and Chemical Equilibrium

As discussed in Chapter 12, chemical reactions are reversible processes that tend to proceed in a direction that approaches chemical equilibrium. Although the equilibrium constant for a given reaction aids considerably in predicting the spontaneous direction of chemical change, it tells us very little about the fundamental driving forces that make chemical reactions spontaneous. In our context, spontaneous reactions are processes that occur naturally without any outside intervention. To explain the spontaneity of chemical reactions, we need to continue our discussion of the thermodynamics involved with chemical change. Although reaction enthalpy is an important consideration, it is not the only factor that governs reaction spontaneity. Every chemical reaction has an accompanying change in the state of disorder or entropy of the system. In this chapter the spontaneity of a chemical reaction is described in terms of changes in enthalpy and entropy that accompany reaction. In addition, a means of predicting the equilibrium constant for any chemical process from thermodynamic considerations is developed.

ESSENTIAL BACKGROUND

- **Logarithms**
- **The first law of thermodynamics and enthalpy of reaction (Chapter 4)**
- **Chemical equilibrium (Chapter 12)**

TOPIC 1: ENTROPY AND SPONTANEOUS PROCESSES

KEY POINTS

✓ *What is entropy a measure of? In what units is it expressed?*

✓ *How is change in entropy related to reaction spontaneity?*

✓ *What is the third law of thermodynamics?*

A chemical or physical change that occurs without outside intervention is called a **spontaneous process**. Examples of spontaneous processes abound. Ice melts to form liquid water at temperatures above 0°C. A skydiver plummets Earthward when jumping out of a flying aircraft. A steel shovel rusts when exposed to oxygen and moisture. Although these spontaneous processes obviously differ in many ways, accompanying each is a net increase in the disorder or entropy of the universe. **Entropy (S)** is a thermodynamic property that quantifies the extent of randomness of a system and is typically expressed in units of joules per Kelvin (J/K). Similar to internal energy and enthalpy, entropy is a state function. Therefore, the change in entropy depends only on initial and final states accessed: $\Delta S = S_{final} - S_{initial}$. A positive value of ΔS indicates an increase in randomness and a negative value indicates a decrease in randomness.

On a molecular level, entropy is related to the number of possible arrangements of energy and position available for a given system. Systems with a large number of possible arrangements are statistically more likely to occur and therefore are entropically favored. For example, consider the sublimation of a mole of solid carbon dioxide: $CO_{2(s)} \rightarrow CO_{2(g)}$. In solid CO_2, molecules are held close together in a crystalline lattice and possess relatively few possible spatial arrangements. In contrast, CO_2 molecules in the gas phase are continuously reordering themselves in an enormous number of possible arrangements in space. Therefore, an increase in entropy is expected to accompany sublimation. From our simple illustration, it follows that for a given material the entropy associated with the three states of matter can be ranked $S_{solid} < S_{liquid} < S_{gas}$.

The **second law of thermodynamics** specifies that every spontaneous process is accompanied by an increase in the entropy of the universe. It follows that $\Delta S_{univ} > 0$ for spontaneous processes and $\Delta S_{univ} < 0$ for nonspontaneous processes. The change in entropy of the universe can be conveniently represented by the sum of contributions from the system and the surroundings: $\Delta S_{univ} = \Delta S_{sys} + \Delta S_{surr}$. The entropy change of the system is related to the number of arrangements of energy and position available to reactant and product molecules. The entropy change of the surroundings is primarily defined by the flow of heat into or out of the system. A flow of heat out of the system leads to a net increase in the random motions of the molecules comprising the surroundings and results in a positive value of ΔS_{surr}. Therefore, reaction exothermicity constitutes an important driving force for spontaneous change. In addition to reaction enthalpy, the magnitude of ΔS_{surr} also depends inversely on the temperature at which heat is exchanged. The dependence of ΔS_{surr} on both variables is expressed mathematically by the equation $\Delta S_{surr} = -\Delta H_{sys} / T$. Substituting this definition into the definition of ΔS_{univ} yields $\Delta S_{univ} = \Delta S_{sys} - \Delta H_{sys} / T$. From this equation it is clear that both changes in enthalpy and entropy contribute in determining reaction spontaneity.

Entropy differs considerably from the other thermodynamic state functions we have discussed in that it can be calculated absolutely. This characteristic of entropy arises from the **third law of thermodynamics** that states that *the entropy of a perfect crystal at absolute zero is equal to zero*. The entropy of a material at standard state conditions is called the **standard entropy ($S°$)**. $S°$ values are known for most compounds. Therefore, the entropy change of the system accompanying any chemical reaction can be calculated by subtracting the standard entropy values of the reactants from those of the products: $\Delta S_{sys} = \Sigma n_p S°_{products} - \Sigma n_r S°_{reactants}$. In this expression, n represents the stoichiometric coefficients in the balanced equation.

Topic Test 1: Entropy and Spontaneous Processes

True/False

1. The entropy of the system increases when water vapor is condensed.

2. Spontaneous processes always increase the entropy of the universe.

3. Entropy, enthalpy, and internal energy are all state functions.

Multiple Choice

4. The standard state enthalpies of gaseous CH_4, O_2, CO_2, and H_2O are 186, 205, 214, and 189 J/(mol K), respectively. Calculate the standard state entropy change associated with the following reaction: $CH_{4(g)} + 2O_{2(g)} \rightarrow CO_{2(g)} + 2H_2O_{(g)}$.
   a. 12 J/(mol K)
   b. −12 J/(mol K)
   c. −4 J/(mol K)
   d. 4 L/mol
   e. None of the above

5. Calculate the ΔS_{surr} when 2.00 moles of liquid water are vaporized at 100°C and 1 atm pressure. $H_2O_{(l)} \rightarrow H_2O_{(g)}$ $\Delta H = 40.6$ kJ/mol.
   a. −218 J/K
   b. 81.2 kJ/K
   c. −109 J/K
   d. 813 J/K
   e. −813 J/K

Short Answer

6. The condensation of a material is characterized by $\Delta S_{surr} = 350.0$ J/K and $\Delta S_{sys} = -31$ J/K at 20°C. Will this process be spontaneous or nonspontaneous?

Topic Test 1: Answers

1. **False.** The entropy change accompanying condensation is less than zero ($\Delta S < 0$). This is because molecules in the gas phase exist in a more disordered state than in the liquid phase.

2. **True.** This is a restatement of the second law of thermodynamics.

3. **True.** Entropy, enthalpy, and internal energy are all state functions, which means they depend only on initial and final states.

4. **c.** The standard state change in entropy is calculated by subtracting the sum of reactant entropies from product entropies. $\Delta S_{sys} = \Sigma n_p S^\circ_{products} - \Sigma n_r S^\circ_{reactants}$.

$$\Delta S_{sys} = [2(189 \text{ J}/(\text{K mol})) + 214 \text{ J}/(\text{K mol})] - [186 \text{ J}/(\text{K mol}) + 2(205 \text{ J}/(\text{K mol}))]$$
$$= -4 \text{ J}/(\text{mol K})$$

5. **a.** The change in the entropy of surroundings can be related to the reaction enthalpy and temperature. Remember to account for the fact that 2 moles of water undergo reaction.

$$\Delta S_{surr} = -\Delta H/T = -[(2 \text{ mol})(40.67 \text{ kJ}/\text{mol})(1,000 \text{ J}/\text{kJ})]/(373 \text{ K}) = -218 \text{ J}/\text{K}$$

6. Spontaneous. The condensation will be spontaneous if $\Delta S_{univ} > 0$. The entropy change of the universe is calculated by taking the sum of contributions from the system and surroundings.

$$\Delta S_{univ} = \Delta S_{sys} + \Delta S_{surr} = -31 \text{ J}/\text{K} + 350 \text{ J}/\text{K} = 319 \text{ J}/\text{K}$$

TOPIC 2: GIBBS FREE ENERGY

KEY POINTS

✓ *What is Gibbs free energy?*

✓ *How can free energy be used to predict reaction spontaneity?*

✓ *How does free energy depend on temperature?*

According to the second law of thermodynamics, the sign of ΔS_{univ} determines whether a particular process is spontaneous. ΔS_{univ} in turn depends on several variables, including ΔS_{sys}, ΔH_{sys}, and temperature. These three variables can be combined in the form of a new state function, **Gibbs free energy (G)**, which also predicts the spontaneity of a chemical or physical process. **Gibbs free energy** is defined by the equation $G = H - TS$, where H is enthalpy, T is temperature, and S is entropy. Generally, chemists are interested in the change in free energy (ΔG) that accompanies chemical reaction: $\Delta G = \Delta H - T\Delta S$. The sign of ΔG indicates the spontaneity of a chemical reaction occurring under conditions of constant temperature and pressure. Under these conditions, ΔG is negative for spontaneous processes and ΔG is positive for nonspontaneous processes. The change in free energy for a system at equilibrium is equal to zero. As shown in the definition of ΔG, when ΔS and ΔH possess the same sign reaction, spontaneity depends on temperature. Under these circumstances, the effect of ΔS will dominate at high temperatures and the influence of ΔH will dominate at low temperatures. The thermodynamic considerations used for predicting reaction spontaneity are summarized in **Table 13.1**.

The **standard free energy change (ΔG^o)** is the change in free energy that occurs when reactants in their standard states are converted to products in their standard states. ΔG^o is usually expressed in terms of the standard state enthalpy and entropy of a given reaction: $\Delta G^o = \Delta H^o - T\Delta S^o$. The **standard free energy of formation (ΔG_f^o)** is defined as the free energy change that occurs when 1 mole of a species is formed from its elements under standard state conditions in their most stable forms. For example, the standard free energy change for the formation of benzene (C_6H_6) is the free energy change associated with the reaction of 1 atm H_2 with pure solid carbon, $6C_{(s)} + 3H_{2(g)} \rightarrow C_6H_{6(l)}$, and is equal to 124.5 kJ/mol. ΔG_f^o values are available for most compounds and are equal to zero for elements in their most stable state. Because free energy is a state function, the ΔG^o for any chemical reaction can be expressed in terms of the standard free energies of formation of all product and reactant species via the equation: $\Delta G^o = \Sigma n_p \Delta G_{f(products)}^o - \Sigma n_r \Delta G_{f(reactants)}^o$. For example, the standard free energy change for the reaction $H_{2(g)} + CO_{2(g)} \rightarrow H_2CO_{2(g)}$ is determined in the following calculation:

$$\Delta G^o = \Sigma n_r \Delta G_{f(products)}^o - \Sigma n_r \Delta G_{f(reactants)}^o = \left[\Delta G_f^o(H_2CO_2)\right] - \left[\Delta G_f^o(H_2) + \Delta G_f^o(CO_2)\right]$$
$$= (-351 \text{kJ/mol}) - (0 + -394 \text{kJ/mol}) = \textbf{43 kJ/mol}$$

Table 13.1 Thermodynamic Considerations to Predict Reaction Spontaneity		
ΔH	ΔS	REACTION SPONTANEITY
<0	>0	Spontaneous at all temperatures
>0	<0	Nonspontaneous at all temperatures
>0	>0	Spontaneous at high temperatures
<0	<0	Spontaneous at low temperatures

Topic Test 2: Gibbs Free Energy

True/False

1. A reaction in which ΔS and ΔH are both positive will tend to be spontaneous at high temperatures.

2. The ΔG_f° of O_2 is equal to zero at 298 K and 1 atm.

3. ΔG for a system at equilibrium is always less than 0.

Multiple Choice

4. Given that ΔH° equals -91.6 kJ/mol and ΔS° equals -197 J/(mol K), calculate the standard free energy change for the following reaction at 25°C: $N_{2(g)} + 3H_{2(g)} \rightarrow 2NH_{3(g)}$.
   a. 150 kJ/mol
   b. 58,600 kJ/mol
   c. -86.7 kJ/mol
   d. -32.9 kJ/mol
   e. 32.9 kJ/mol

5. ΔH is equal to 7.10 kJ and ΔS is equal to 33.0 J/K for a given chemical reaction. Above what temperature is the reaction expected to proceed spontaneously?
   a. 1,510 K
   b. 298 K
   c. 215 K
   d. Spontaneous at all temperatures
   e. Nonspontaneous at all temperatures

Short Answer

6. The standard free energy of formations of $C_2H_{2(g)}$ and $CO_{2(g)}$ are 209 kJ/mol and -394.4 kJ/mol, respectively. Calculate the standard free energy of formation for $H_2O_{(g)}$ given the following thermodynamic equation: $C_2H_{2(g)} + 3O_{2(g)} \rightarrow 2CO_{2(g)} + 2H_2O_{(g)}$ $\Delta G^\circ = -1,455$ kJ/mol.

Topic Test 2: Answers

1. **True.** At high temperatures, free energy is dominated by the contribution from entropy. Therefore, ΔG will be negative and the process will be spontaneous.

2. **True.** ΔG_f° is equal to zero for elements in their most stable state under standard state conditions.

3. **False.** ΔG is equal to zero for a system at equilibrium.

4. **d.** The free energy change for a reaction can be calculated given the changes in enthalpy and entropy change. $\Delta G^\circ = \Delta H^\circ - T\Delta S^\circ = -91.6$ kJ $- [(298\,\text{K})\,(-197\,\text{J/K})\,(1\,\text{kJ}/1,000\,\text{J})] =$ **-32.9 kJ/mol**.

5. **c.** Because ΔH and ΔS possess different signs, the spontaneity of the reaction will depend on temperature. At high temperatures, the reaction is expected to be spontaneous. Recall that for a spontaneous reaction: $\Delta G = \Delta H - T\Delta S < 0$.

Rearranging the equation and solving:

$$T > \Delta H / \Delta S = [(7.10\,kJ)(1{,}000\,J/1\,kJ)]/(33.0\,J/K) = \mathbf{215K}$$

6. $-228.6\,kJ/mol$. The standard free energy change for a reaction may be expressed in terms of the standard free energies of formation for reactants and products: $\Delta G^\circ = Sn_p \Delta G^\circ_{f(products)} - Sn_r \Delta G^\circ_{f(reactants)}$.

Inserting values of ΔG°_f and rearranging the equation:

$$2 \times \Delta G^\circ_f(H_2O) = \Delta G^\circ - 2 \times \Delta G^\circ_f(CO_2) + \Delta G^\circ_f(C_2H_2)$$
$$= -1{,}455\,kJ/mol - 2(-394.4\,kJ/mol) + (209\,kJ/mol) = -457.2\,kJ/mol$$

Solving for ΔG°_f (H_2O):

$$\Delta G^\circ_f(H_2O) = (-457.2\,kJ/mol)/2 = \mathbf{-228.6\,kJ/mol}$$

TOPIC 3: FREE ENERGY AND CHEMICAL EQUILIBRIUM

KEY POINTS

✓ *How is the change in free energy for non-standard state conditions determined?*

✓ *What is the relationship between free energy and chemical equilibrium?*

✓ *How does free energy relate to work?*

Nature reacts spontaneously to achieve the lowest possible state of free energy. The direction of chemical change that is associated with decreasing free energy can be predicted for standard state conditions given the standard free energy change for a reaction. However, does this imply that a reaction will proceed in the spontaneous direction until all reactants are consumed? Actually, a state always exists even lower in free energy than that represented by the products of a chemical reaction. This state is the minimum free energy available to the system and reflects a condition of chemical equilibrium. For example, if 1 mole of N_2 is placed in a vessel containing 3 moles of H_2, the following reversable reaction will occur spontaneously: $N_{2(g)} + 3H_{2(g)} \rightarrow 2NH_{3(g)}$. As the reaction proceeds, the concentration of NH_3 increases and the free energy of the system decreases. Eventually, the reaction will stop short of going to completion and the concentration of NH_3 will remain constant. At this point the system will be in equilibrium and exist in the lowest possible free energy state.

As just noted, the free energy of a system changes as a reaction progresses. Therefore, chemists need to be able to describe the change in free energy when reactants and products are in non-standard states. This free energy is designated by the symbol $\mathbf{\Delta G}$ without the standard state symbol (°) and is calculated using the expression: $\mathbf{\Delta G = \Delta G^\circ + RT\ln Q}$, where Q is equal to the reaction quotient. Q is defined in an identical fashion as the equilibrium constant but reflects reactant and product concentrations for nonequilibrium conditions. The expression above also relates the value of the equilibrium constant to the standard free energy change of a process. At equilibrium, the free energy of a system ceases to change because it has achieved the lowest possible free energy state. Therefore, $\Delta G = 0$ and $Q = K$ for a system at equilibrium. Substituting these values into the equation above yields the expression: $\mathbf{\Delta G^\circ = -RT\ln K}$. Therefore, the equilibrium constant for any chemical reaction can be calculated given the standard free energy change. It is important to note that equilibrium constants calculated in this way are in molarity units for solution-phase reactions and in pressure units for gas-phase processes.

In addition to predicting the direction of spontaneous change, free energy provides useful information pertaining to the amount of energy available from a chemical process to do work. Indeed, the maximum amount of work obtainable from a given spontaneous process occurring at a constant pressure and temperature is equal to ΔG. This relationship is expressed mathematically by $\mathbf{w_{max}} = \mathbf{\Delta G}$, where w_{max} represents the maximum amount of work. Similarly, the value of ΔG for processes that are not spontaneous reflects a measure of the minimum amount of work that must be expended to cause the process to occur: $\mathbf{w_{min}} = \mathbf{\Delta G}$. In reality, the values of w_{max} and w_{min} represent theoretical limits due to the inefficiencies in the manner in which work is performed.

Topic Test 3: Free Energy and Chemical Equilibrium

True/False

1. ΔG increases as a system approaches equilibrium.

2. Equilibrium is the point of lowest free energy available to a system.

3. For a spontaneous process, ΔG is equal to the amount of work that must be expended to cause the process to occur.

Multiple Choice

4. Calculate the equilibrium constant for the following reaction at 25°C given that the $\Delta G_f^\circ (NO_{2(g)}) = 51\,kJ/mol$. $2O_{2(g)} + N_{2(g)} \rightleftarrows 2NO_{2(g)}$
   a. $2.56 \times 10^{13}\,atm^{-1}$
   b. $8.84 \times 10^{8}\,atm^{-1}$
   c. $1.28 \times 10^{-18}\,atm^{-1}$
   d. $7.82 \times 10^{17}\,atm^{-1}$
   e. None of the above

5. The K_{sp} for the dissolution of $NaCl_{(s)}$ is equal to 41.1 at 25°C. Calculate the standard free energy change associated with this process.
   a. $-9.21\,kJ/mol$
   b. $-10.2\,kJ/mol$
   c. $-0.772\,kJ/mol$
   d. $0.772\,kJ/mol$
   e. $-0.0309\,kJ/mol$

Short Answer

6. Given the following free energies of formation, what is the maximum amount of work obtainable from the combustion of 2 moles of ethane under standard state conditions?

$$2C_2H_{6(g)} + 7O_{2(g)} \rightarrow 4CO_{2(g)} + 6H_2O_{(g)}$$
$$\Delta G_f^\circ (C_2H_{6(g)}) = -32.9\,kJ/mol$$
$$\Delta G_f^\circ (CO_{2(g)}) = -394\,kJ/mol$$
$$\Delta G_f^\circ (H_2O_{(g)}) = -228.6\,kJ/mol$$

Topic Test 3: Answers

1. **False.** ΔG decreases as a system approaches equilibrium.

2. **True.** The free energy of a system at equilibrium is the lowest free energy state available to the system.

3. **False.** ΔG is equal to the maximum work obtainable from a spontaneous process.

4. **c.** The standard free energy change for a reaction is related to the equilibrium constant by the expression $\Delta G° = -RT \ln K$. To obtain $\Delta G°_{rxn}$, double the value of $\Delta G°_f$.

   $$\ln K = -\Delta G°/RT = [(2 \times 51\,kJ/mol)(1,000\,J/1\,kJ)]/[(8.3145\,J/(K\,mol))(298\,K)] = -41.2$$

   $$K = 1.28 \times 10^{-18}\,atm^{-1}.$$

5. **a.** The dissolution follows the reaction $NaCl_s \rightleftarrows Na^+_{(aq)} + Cl^-_{(aq)}$. $\Delta G°$ is determined by the following calculation: $\Delta G° = -RT \ln K = -(8.3145\,J/K)(1\,kJ/1,000\,J)(298\,K)\ln(41.1)$
   $= -9.21\,kJ/mol$.

6. **−2,882 kJ.** The maximum amount of work obtainable from combustion will equal the free energy change for the process $w_{max} = \Delta G_{rxn}$. The free energy change for ethane combustion is calculated by subtracting the standard free energies of formation of reactants from those of the products. Remember that $\Delta G°_f = 0$ for $O_{2(g)}$.

   $$\Delta G° = \Sigma n_p \Delta G°_{f(products)} - \Sigma n_r \Delta G°_{f(reactants)}$$
   $$= [(6 \times \Delta G°_f(H_2O)) + (4 \times \Delta G°_f(CO_2))] - [2 \times \Delta G°_f(C_2H_6)]$$
   $$= 6 \times (-228.6\,kJ/mol) + 4 \times (-394\,kJ/mol) - 2 \times (-32.9\,kJ/mol) = -2,882\,kJ.$$

APPLICATION

Mitochondria are the power plants of most eukaryotic cells. These cytoplasmic organelles are ellipsoids approximately 0.5 mm in diameter and consist of an outer membrane, an extensively coiled intermembrane, and a gel-like intercompartment called the matrix. Mitochondria comprise roughly one fifth of the total volume of a cell and are responsible for producing the energy required to perform most cellular functions. These functions include the synthesis of proteins and DNA and maintenance of homeostasis. Because most "cellular work" is thermodynamically unfavorable ($\Delta G > 0$), chemical reactions of compounds manufactured in the mitochondria are coupled to these processes to overcome the thermodynamic barriers associated with them. In this way, the cell is able to do work and perform the necessary biological tasks to sustain an organism. The mechanisms by which cellular mitochondria provide energy to a eukaryotic cell couples the oxidation of nutrients to the production of adenosine triphosphate (ATP). These enzyme-catalyzed reactions occur in the matrix and on the surface of the inner mitochondrial membrane. Upon production and release to the rest of the cell, ATP fuels the various energy-consuming processes comprising cellular work. One way that ATP generates energy is via hydrolysis in the presence of an enzyme to produce adenosine diphosphate (ADP): ATP + H_2O → ADP + phosphate ion; $\Delta G = -30\,kJ$. By coupling the spontaneous hydrolysis of ATP to important nonspontaneous chemical reactions, cells are able to sustain themselves and multiply.

DEMONSTRATION PROBLEM

Approximately 30 billion pounds of ammonia (NH_3) is manufactured annually for use primarily in the production in chemical fertilizers. Typically, this compound is synthesized by the Haber process involving the reaction of nitrogen and hydrogen gases: $N_{2(g)} + 3H_{2(g)} \leftrightarrow 2NH_{3(g)}$. If the standard enthalpy (ΔH_f°) and entropy (ΔS_f°) of formation of ammonia are $-46\,kJ/mol$ and $193\,J/(mol\,K)$, respectively, calculate the equilibrium constant for the Haber process reaction at $25°C$.

Solution

Calculate the Gibbs free energy of formation (ΔG_f°) for $NH_{3(g)}$:

$\Delta G_f^\circ = \Delta H_f^\circ - T\Delta S_f^\circ = -46\,kJ/mol - (298.15\,K) \times (193\,J/(mol\,K)) \times (1\,kJ/1000\,J)$

$= -104\,kJ/mol$

Calculate the change in Gibbs free energy (ΔG°) for the Haber reaction:

$\Delta G_{rxn}^\circ = \left[2 \times \Delta G_f^\circ(NH_3)\right] - \left[3 \times \Delta G_f^\circ(H_2) + \Delta G_f^\circ(N_2)\right] = -208\,kJ/mol$

Calculate the equilibrium constant (K) for the Haber reaction:

$\Delta G_{rxn}^\circ = -RT\ln(K)$

Rearranging the equation and solving for K:

$K = \exp\left[-\Delta G_{rxn}^\circ/RT\right] = \exp[((208\,kJ/mol) \times (1000\,J/1\,kJ))/((8.314\,J/(K\,mol) \times 298.15\,K))]$

$K = \exp(83.9) = 2.7 \times 10^{36}$

Chapter Test
True/False

1. Spontaneous processes can be used to perform work.

2. A reaction with $\Delta S > 0$ and $\Delta H > 0$ will be spontaneous at every temperature.

3. The entropy of 1 mole of ice is greater than that of 1 mole of water vapor.

4. A system in chemical equilibrium has a $\Delta G = 0$.

5. The entropy of the universe is a constant ($\Delta S_{universe} = 0$).

Multiple Choice

6. Which of the following reactions is expected to have a negative change in entropy?
   a. $H_2O_{(l)} \rightarrow H_2O_{(g)}$
   b. $H_2O_{(s)} \rightarrow H_2O_{(g)}$
   c. $H_2O_{(s)} \rightarrow H_2O_{(l)}$
   d. $CaCO_{3(s)} \rightarrow CaO_{(s)} + CO_{2(g)}$
   e. $CaO_{(s)} + CO_{2(g)} \rightarrow CaCO_{3(s)}$

7. Given that ΔH° equals $-91.8\,kJ$ and ΔG° equals $-33.1\,kJ$ at $298\,K$, what is the standard state change in entropy (ΔS°) for the reaction $N_{2(g)} + 3H_{2(g)} \rightarrow 2\,NH_{3(g)}$.

a. $-197 \, \text{J/K}$
b. $-150 \, \text{J/K}$
c. $-58.7 \, \text{J/K}$
d. $58.7 \, \text{J/K}$
e. $-125 \, \text{J/K}$

8. Ammonium and chloride ions react according to the following equation: $NH_{4(aq)}^+ + Cl_{(aq)}^- \rightleftharpoons NH_4Cl_{(s)}$; $\Delta G^\circ = 7.0 \, \text{kJ}$. What is the equilibrium constant for this reaction at 25°C?
   a. $1.0 \, \text{L}^2/\text{mol}^2$
   b. $4.5 \times 10^3 \, \text{L}^2/\text{mol}^2$
   c. $5.9 \times 10^{-2} \, \text{L}^2/\text{mol}^2$
   d. $1.1 \times 10^{-4} \, \text{L}^2/\text{mol}^2$
   e. None of the above

9. Consider the following thermochemical equation at 25°C and 1 atm pressure: $CH_{4(g)} + N_{2(g)} \rightarrow HCN_{(g)} + NH_{3(g)}$; $\Delta H^\circ = 164 \, \text{kJ}$ and $\Delta S^\circ = 20.1 \, \text{J/K}$. Under what conditions will this reaction proceed spontaneously?
   a. Low temperatures
   b. High temperatures
   c. All temperatures
   d. The reaction will never be spontaneous

10. Consider the standard state entropy change of the following reaction: $CCl_{4(g)} + O_{2(g)} \rightarrow CO_{2(g)} + 2Cl_{2(g)}$; $\Delta S^\circ = 239 \, \text{J/K}$. Given that the standard state entropies (S°) for $CCl_{4(g)}$, $CO_{2(g)}$, and $Cl_{2(g)}$ are 216, 214, and 223 J/K, respectively, what is the standard state entropy of $O_{2(g)}$?
    a. 0
    b. $205 \, \text{J/K}$
    c. $16 \, \text{J/K}$
    d. $121 \, \text{J/K}$
    e. $-16 \, \text{J/K}$

Short Answer

11. Below what temperature will the following reaction be spontaneous?

    $NO_{2(g)} + NO_{2(g)} \rightarrow N_2O_{4(g)}$; $\Delta S^\circ = -176 \, \text{J/K}$ and $\Delta H^\circ = -58.0 \, \text{kJ}$

12. Ethylamine ($C_2H_5NH_2$; $K_b = 4.38 \times 10^{-4}$) is a weak base emitted by high-altitude wetlands. What is the standard free energy change (ΔG°) associated with the following reaction at 298 K?

    $C_2H_5NH_{2(g)} + H_2O_{(g)} \rightarrow C_2H_5NH_{(g)}^+ + OH_{(g)}^-$.

13. The standard state enthalpies (S°) of $NH_{3(g)}$, $O_{2(g)}$, $H_2O_{(g)}$, and $NO_{2(g)}$ are 193, 205, 70.0, and 240 J/K, respectively. Calculate ΔS° for the reaction $4NH_{3(g)} + 7O_{2(g)} \rightarrow 4NO_{2(g)} + 6H_2O_{(l)}$.

14. What is the entropy of a perfect crystal of $CaCO_3$ at zero K.

15. What is the equilibrium constant (K_{sp}) for the dissolution of calcium fluoride at 1,114 K: $CaF_{2(s)} \rightleftharpoons Ca_{(aq)}^{2+} + 2F^-$; $\Delta G^\circ = -59.3 \, \text{kJ}$.

16. Given the following thermochemical reaction, $2Ag_{(s)} + Cl_{2(g)} \rightarrow 2AgCl_{2(s)}$ $\Delta G° = -210\,kJ$, calculate the free energy change (ΔG) for the reaction of $0.231\,atm$ Cl_2 and 175 grams of $Ag_{(s)}$ at $298\,K$.

Chapter Test Answers

1. **True**
2. **False**
3. **False**
4. **True**
5. **False**
6. **e** 7. **a** 8. **c** 9. **b** 10. **b**
11. $329\,K$
12. $19.2\,kJ$
13. $-827\,J/K$
14. $0\,J/K$
15. 603
16. $-206\,kJ$

Check Your Performance

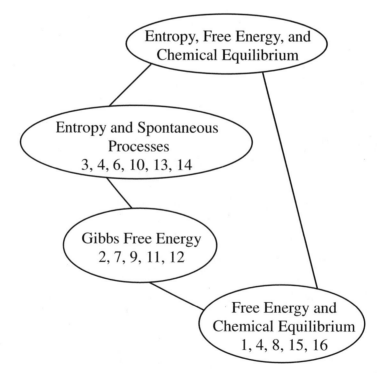

Use this chart to identify weak areas, based on the question numbers you answered incorrectly in the chapter test.

Final Exam

True/False

1. 4.0 moles of gaseous HCl will occupy a volume of 89.6 liters at 0°C and 1 atm.

2. The addition of a solute *always* decreases the vapor pressure of a solvent.

3. The free energy of a system at equilibrium is equal to zero.

4. During a spontaneous process entropy is always conserved.

5. A catalyst increases reaction rate without being consumed during the course of a reaction.

Multiple Choice

6. The average spring time loss rate of ozone (O_3) in the Antarctic stratosphere is measured to 9.0×10^{-3} atm sec^{-1}. Given that the reaction responsible for the O_3 destruction is $2O_3 \rightarrow 3O_2$, what is the rate of appearance of O_2 generated in this reaction?
   a. 9.0×10^{-3} atm sec^{-1}
   b. 1.4×10^{-2} atm sec^{-1}
   c. 6.0×10^{-3} atm sec^{-1}
   d. 3.0×10^{-5} atm sec^{-1}
   e. 2.7×10^{-5} atm sec^{-1}

7. In the following reaction mechanism responsible for the destruction of O_3 in the Arctic troposphere, which species acts as a catalyst in the mechanism?

$$Br + O_3 \rightarrow BrO + O_2$$

$$BrO + O_3 \rightarrow Br + 2O_2$$

   a. Br
   b. O_3
   c. BrO
   d. O_2

8. Given the rate constant data below, calculate the activation energy for the reaction $N_2O_5 \rightarrow NO_2 + NO_3$.

Temperature (K)	Rate Constant (Lmol^{-1}sec^{-1})
298	0.113
450	0.759

   a. 14 kJ/mol
   b. 25 kJ/mol
   c. 120 J/mol
   d. 3,500 J/mol
   e. 9.0 kJ/mol

9. A 10.0-gram sample of solid NH_4Cl is placed in an evacuated reaction vessel (10 liters) at 1,110 K. At equilibrium the total pressure of the reactor is equal to 2.40 atm.

$$NH_4Cl_{(s)} \leftrightarrows NH_{3(g)} + HCl_{(g)}$$

What is the equilibrium constant (K_p) for this reaction?

    a. 2.40 atm
    b. 1.44 atm
    c. 4.80 atm
    d. 101 atm
    e. None of the above

10. Which of the following compounds is the strongest acid?

    a. HF
    b. H_2O
    c. H_2SO_4
    d. NH_4^+
    e. CH_3OH

11. What is the pH of a 5.5 M solution of pyridine (C_5H_5N; $K_b = 1.7 \times 10^{-9}$)?

    a. 2.36
    b. −0.070
    c. 9.30
    d. 10
    e. 12

12. Water spontaneously dissociates to form H^+ and OH^- ions; $H_2O_{(l)} \rightleftharpoons H^+_{(aq)} + OH^-_{(aq)}$. At 0°C, the auto-ionization constant of water is equal to 1.2×10^{-15}. What is the pH of water at 0°C?

    a. 7.46
    b. 9.88
    c. 7.56
    d. 4.12
    e. None of the above

13. Calculate the ΔG°_{rxn} for the combustion of 2 moles of C_2H_5OH at 25°C using the standard free energies of formation given below.

$$C_2H_5OH + 3O_2 \rightarrow 2CO_2 + 3H_2O$$

$$\Delta G^\circ_f \text{ (kJ/mol)}$$

C_2H_5OH	−174.8
CO_2	−394.4
H_2O	−228.6

    a. 723.1 kJ
    b. 1,865 kJ
    c. −1,299 kJ
    d. 991.1 kJ
    e. −2,599.6 kJ

14. The molar solubility of lead chloride ($PbCl_{2(s)}$) is equal to 1.6×10^{-2} M at 25°C. What is the solubility product (K_{sp}) of $PbCl_2$?

    a. 1.6×10^{-2}
    b. 4.1×10^{-6}

c. 7.1×10^{-5}
d. 1.6×10^{-5}
e. 9.3×10^{-3}

15. A $1.12\,M$ solution of HF has a pH of 1.55 at 25°C. What is the standard free energy change (ΔG°) for the dissociation reaction of HF to form H^+ and F^- ions?
    a. 18 kJ
    b. −30 kJ
    c. 0 kJ
    d. 16 kJ
    e. 61 kJ

16. What is the density (g/L) of a gaseous sample of CO_2 at standard temperature and pressure? (Hint: assume it behaves as an ideal gas.)
    a. $1.8234 \times 10^3\,g/L$
    b. $7.9234 \times 10^{-2}\,g/L$
    c. $7.9234\,g/L$
    d. $4.2671 \times 10^{-5}\,g/L$
    e. $1.9634\,g/L$

17. Assuming ideal behavior, what volume does a 42.0-gram sample of Ar occupy at 0.1789 atm and 214.1 K?
    a. 620 liters
    b. 412 liters
    c. 103 liters
    d. 1,130 liters
    e. None of the above

18. Which of the following materials is expected to have the lowest boiling point temperature?
    a. HF
    b. H_2O
    c. HCl
    d. HNO_3
    e. CF_4

Short Answer

19. Acetone, CH_3COCH_3, is a volatile organic compound emitted by growing vegetation. Acetone has a ΔH_{vap} equal to $40\,kJ/mol$ and a standard boiling temperature of 330 K. What is the vapor pressure of this compound at room temperature or 298 K?

20. The use of road salt to prevent icy streets takes advantage of the freezing point depression of saline solutions. Calculate the freezing point temperature of a $0.550\,M$ solution of NaCl that has a density equal to $1.07\,g/mL$. [k_f (H_2O) $= 1.86°C\,kg/mol$].

21. Many of the important hormones involved in the complex chemical signaling in the body act as weak acids and bases. For example, adrenaline ($C_9O_3H_{12}NH$, $K_b = 1.23 \times 10^{-6}$) is a hormone that is also a weak base. In blood of pH 7.34, what is the ratio of acidic to basic forms of adrenaline ($[C_9O_3H_{12}NH_2^+]/[C_9O_3H_{12}NH]$) present?

22. A 33.0-gram sample of $MgCO_3$ is placed in a 33.1-liter evacuated cylinder and heated to 1,001 K. Assuming $MgCO_3$ decomposes via the following reaction, what is the pressure of the cylinder upon complete reaction?

$$MgCO_{3(s)} \rightarrow MgO_{(s)} + CO_{2(g)}$$

23. What is the pH of a solution made by mixing 151 milliliters of 0.152 M HCl with 201 milliliters of 0.122 M NaOH?

24. Calculate the boiling point of a solution prepared by dissolving 33.1 grams of glucose ($C_6H_{12}O_6$, nonelectrolyte) in 252 grams of water. ($K_b = 0.510°C\,kg/mol$)

25. Calculate the partial pressure of acetone at 25°C in a solution prepared by mixing 44.1 grams of acetone (C_3H_6O, volatile liquid) with 51.2 grams of benzene (C_6H_6, volatile liquid). The vapor pressures of pure acetone and benzene at 25°C are 251 and 94.4 Torr, respectively.

26. A drug is removed from the body by a first-order loss process with a rate coefficient of $4.132 \times 10^{-4}\,sec^{-1}$. How many days will it take a patient's body to remove 99% of the drug upon stopping the medication?

27. Dilute solutions of H_2O_2 (M.M. = $34.0\,g\,mol^{-1}$) are widely used as a medical sterilization agent in most hospitals. Calculate the molarity of a 30% by mass solution of H_2O_2 that has a density of $1.11\,g\,cm^{-3}$.

28. Carbon monoxide (CO) is a toxic product of the incomplete combustion of fossil fuels. CO is oxidized in the atmosphere by reaction with OH:

$$CO + OH \rightarrow CO_2 + H$$

If the A factor of this reaction is $1.50 \times 10^{-13}\,cm^3\,mol^{-1}\,sec^{-1}$ and the activation energy is equal to $1,000.0\,cal\,mol^{-1}$, what is the rate coefficient for this reaction at 1,501°C?

29. The world's oceans constitute an enormous sink for CO_2 generated in the combustion of fossil fuels. Calculate the concentration of CO_2 expected at the ocean's surface if the partial pressure of CO_2 equals 19.06 Pa and the Henry's law coefficient for CO_2 in seawater is equal to $32\,L\,atm\,mol^{-1}$.

30. Hydrofluoric acid is a particularly nasty compound that efficiently dissolves bone material. A laboratory study shows that in a 12 M HF solution 0.77% of the HF molecules are in their dissociated form at 25°C: $HF_{(l)} \leftrightarrows H^+_{(aq)} + F^-_{(aq)}$. What is the equilibrium constant for this process at a temperature of 25°C?

Answers

1. **True**

2. **False**

3. **False**

4. **False**

5. **True**

6. **b**   7. **a**   8. **a**   9. **b**   10. **e**   11. **d**   12. **a**   13. **e**   14. **d**   15. **a**   16. **e**

17. **c**   18. **e**

19. 160 Torr

20. $-0.956°C$

21. 5.59

22. 0.971

23. 11.196

24. $100.37°C$

25. 135 Torr

26. 0.13 days

27. 9.79 M

28. $1.13 \times 10^{-13} \, cm^3/(molecule \; sec)$

29. $5.9 \times 10^{-6} \, M$

30. $K_c = 7.2 \times 10^{-4} \, M$

INDEX